CAMBRIDGE TRACTS IN MATHEMATICS

---

141 Fixed Point Theory and Applications

This book provides a clear exposition of the flourishing field of fixed point theory. Starting from the basics of Banach's contraction theorem, most of the main results and techniques are developed: fixed point results are established for several classes of maps and the three main approaches to establishing continuation principles are presented. The theory is applied to many areas of current interest in analysis. Topological considerations play a crucial role, including a final chapter on the relationship with degree theory. Researchers and graduate students in applicable analysis will find this to be a useful survey of the fundamental principles of the subject. The very extensive bibliography and close to 100 exercises mean that it can be used both as a text and as a comprehensive reference work, currently the only one of its type.

Ravi P. Agarwal
National University of Singapore

Maria Meehan
Dublin City University

Donal O'Regan
National University of Ireland, Galway

Fixed Point Theory and Applications

CAMBRIDGE UNIVERSITY PRESS
Cambridge, New York, Melbourne, Madrid, Cape Town, Singapore, São Paulo, Delhi

Cambridge University Press
The Edinburgh Building, Cambridge CB2 8RU, UK

Published in the United States of America by Cambridge University Press, New York

www.cambridge.org
Information on this title: www.cambridge.org/9780521104197

First published 2001
This digitally printed version 2009

A catalogue record for this publication is available from the British Library

ISBN 978-0-521-80250-5 hardback
ISBN 978-0-521-10419-7 paperback

Contents

Preface

Perhaps the most well known result in the theory of fixed points is Banach's contraction mapping principle. It is therefore fitting that we commence this book with a discussion of contractions and a proof of this result. In addition in Chapter 1, a local version and a generalisation of Banach's contraction theorem are presented. We choose the problem of existence and uniqueness of solutions of certain first order initial value problems to demonstrate the results detailed in the chapter.

It is inevitable that any discussion on contractive maps will lead naturally to another on nonexpansive maps, which is why we choose this as the topic of Chapter 2. Schauder's theorem for nonexpansive maps is presented but the main theorem discussed is a result proved independently in 1965 by Browder, Göhde and Kirk which shows that each nonexpansive map $F : C \to C$, where C is a particular set in a Hilbert space, has at least one fixed point. As a natural lead in to the next chapter, we close Chapter 2 with a nonlinear alternative of Leray–Schauder type for nonexpansive maps.

Chapter 3 is concerned with continuation methods for contractive and nonexpansive maps. We show initially that the property of having a fixed point is invariant by homotopy for contractions. Using this result a nonlinear alternative of Leray–Schauder type is presented for contractive maps and subsequently generalised for nonexpansive maps. An application of the nonlinear alternative for contractions is demonstrated with a second order homogeneous Dirichlet problem.

Fixed point theory for continuous, single valued maps in finite and infinite dimensional Banach spaces is discussed in Chapter 4 with the theorems of Brouwer, Schauder and Mönch presented. The first half of the chapter is devoted to proving and generalising the result of Brouwer which states that every continuous map $F : B^n \to B^n$, where B^n is the closed unit ball in $\mathbf{R}^n$, has a fixed point. An extension of Brouwer's theorem to compact maps in normed linear spaces is presented in the well known fixed point theorem of Schauder, the proof of which relies on Schauder projections. A further generalisation of this theorem due to Mönch in 1980 is also presented. Applications of theorems in Chapter 4 are illustrated with a discrete boundary value problem.

The problem with the results presented in Chapter 4 – from an applications viewpoint – is that they require the map under investigation to take a closed convex set back into itself. As a result, in Chapter 5 we turn our attention to establishing a fixed point theory for nonself maps. However, Schauder's fixed point theorem from Chapter 4 enables us to obtain a nonlinear alternative for continuous, compact (nonself) maps, which we immediately apply to establish an existence principle for a nonlinear Fredholm integral equation. Two existence results are subsequently presented, each ensuring the existence of at least one continuous solution of the equation. In a similar fashion, using Mönch's fixed point theorem from Chapter 4, we obtain a nonlinear alternative for continuous, Mönch type maps which further leads to a variety of nonlinear alternatives for condensing maps, k-set contractive maps, and maps of the form $F := F_1 + F_2$ where F_1 and F_2 satisfy certain conditions. The chapter closes by discussing maps $F : X \to E$ where (unlike the other theorems presented in the chapter) the interior of X may be empty.

Having discussed condensing maps in Chapter 4 it is now natural to consider continuation principles for these maps. There are three main approaches in the literature. The first approach uses degree theory (Chapter 12), the second is the essential map approach of Granas and the third is the 0-epi map approach of Furi, Martelli and Vignoli (Chapter 8). In Chapter 6 we discuss the second approach. The chapter is devoted to showing that the property of having a fixed point (or more generally, being essential) is invariant by homotopy for compact (or more generally, condensing) maps. Using the theory of essential maps, a nonlinear alternative is presented firstly for continuous, compact maps, with an analogous result for continuous, condensing maps appearing towards the end of the chapter. In addition, using results obtained for the completely continuous field, $f(x) = x - F(x)$, the equation $y = x - F(x)$ is discussed, with the Fredholm alternative making a guest appearance as an immediate consequence.

Fixed point theorems in conical shells are discussed in Chapter 7. We consider continuous, compact maps $F : B \to C$, where C is a cone. Specifically we present Krasnoselskii's compression and expansion of a cone theorems, the proofs of which rely on the essential map theory of the previous chapter. Krasnoselskii's theorems are of particular importance when it comes to establishing the existence of multiple solutions of operator equations. A large portion of the chapter is therefore assigned to using these results to prove the existence of multiple solutions of a nonlinear Fredholm integral equation, similar to that discussed already in Chapter 5. In fact, in certain instances we combine the nonlinear alternative for compact maps presented in Chapter 5 and the fixed point theorems of this chapter, to obtain stronger results.

In Chapter 8 we present fixed point results for maps defined on Hausdorff locally convex linear topological spaces. The extension of Schauder's fixed point theorem to such spaces is known as the Schauder–Tychonoff theorem and this is the first main result of the chapter. With this established we then present a nonlinear alternative of Leray–Schauder type for maps defined on the spaces in question. This in turn is used to prove a fixed point theorem of Furi–Pera type which we require to obtain existence principles and results for two types of second order boundary value problems on the half-line. The chapter is concluded with a continuation principle for continuous, compact maps defined on Fréchet spaces. In particular, the 0-epi map approach of Furi, Martelli and Vignoli is presented.

Contractive and nonexpansive multivalued mappings are discussed in Chapter 9. We first concentrate on contractive multivalued mappings. The result due to Nadler of the Banach contraction principle for contractive mappings with closed values is first presented. Once we have shown that the property of having a fixed point is invariant by homotopy for contractive multivalued mappings, a nonlinear alternative of Leray–Schauder type for such maps is then discussed. Finally we extend this result to nonexpansive multivalued mappings.

Chapter 10 deals with multivalued maps which have continuous selection. For the convenience of the reader we restrict our attention to one particular class of maps, namely the $\Phi^\star$ maps. A nonlinear alternative of Leray–Schauder type, a Furi–Pera type result and some coincidence type results are just some of the fixed point theory presented for $\Phi^\star$ maps. An application to abstract economies concludes the discussion.

The objective of Chapter 11 is to extend the Schauder–Tychonoff theorem to multivalued maps with closed graph. The basic results in this chapter are due to S. Kakutani, I. L. Glicksberg and K. Fan. Much of the chapter is spent working towards the proof of Ky Fan's minimax theorem which is the crucial result needed to prove the analogue of the

Schauder–Tychonoff theorem for these maps. The chapter closes with a nonlinear alternative for multivalued maps with closed graph.

A chapter on degree theory concludes the book. The concept of the degree of a map is introduced and is used to provide another approach to presenting fixed point theory and continuation principles. The first half of the chapter deals with the degree of a map defined on subsets of $\mathbf{R}^n$ and a proof of Brouwer's fixed point theorem using the degree theory established is just one of the results presented. The second half of the chapter looks at the degree of a map defined on a normed linear space and here an alternative proof of Schauder's fixed point theorem using degree theory is illustrated.

1

Contractions

Let (X, d) be a metric space. A map $F : X \to X$ is said to be *Lipschitzian* if there exists a constant $\alpha \geq 0$ with

$$d(F(x), F(y)) \leq \alpha \, d(x, y) \text{ for all } x, y \in X. \tag{1.1}$$

Notice that a Lipschitzian map is necessarily continuous. The smallest α for which (1.1) holds is said to be the *Lipschitz constant* for F and is denoted by L. If $L < 1$ we say that F is a *contraction*, whereas if $L = 1$, we say that F is *nonexpansive*.

For notational purposes we define $F^n(x)$, $x \in X$ and $n \in \{0, 1, 2, \ldots\}$, inductively by $F^0(x) = x$ and $F^{n+1}(x) = F(F^n(x))$.

The first result in this chapter is known as Banach's contraction principle.

Theorem 1.1 *Let (X, d) be a complete metric space and let $F : X \to X$ be a contraction with Lipschitzian constant L. Then F has a unique fixed point $u \in X$. Furthermore, for any $x \in X$ we have*

$$\lim_{n \to \infty} F^n(x) = u$$

with

$$d(F^n(x), u) \leq \frac{L^n}{1 - L} \, d(x, F(x)).$$

Proof We first show uniqueness. Suppose there exist $x, y \in X$ with $x = F(x)$ and $y = F(y)$. Then

$$d(x, y) = d(F(x), F(y)) \leq L \, d(x, y),$$

therefore $d(x, y) = 0$.

To show existence select $x \in X$. We first show that $\{F^n(x)\}$ is a Cauchy sequence. Notice for $n \in \{0, 1, \dots\}$ that

$$d(F^n(x), F^{n+1}(x)) \le L\, d(F^{n-1}(x), F^n(x)) \le \dots \le L^n\, d(x, F(x)).$$

Thus for $m > n$ where $n \in \{0, 1, \dots\}$,

$$\begin{aligned} d(F^n(x), F^m(x)) &\le d(F^n(x), F^{n+1}(x)) + d(F^{n+1}(x), F^{n+2}(x)) \\ &\quad + \dots + d(F^{m-1}(x), F^m(x)) \\ &\le L^n\, d(x, F(x)) + \dots + L^{m-1}\, d(x, F(x)) \\ &\le L^n\, d(x, F(x)) \left[1 + L + L^2 + \dots\right] \\ &= \frac{L^n}{1 - L}\, d(x, F(x)). \end{aligned}$$

That is for $m > n$, $n \in \{0, 1, \dots\}$,

$$d(F^n(x), F^m(x)) \le \frac{L^n}{1 - L}\, d(x, F(x)). \tag{1.2}$$

This shows that $\{F^n(x)\}$ is a Cauchy sequence and since X is complete there exists $u \in X$ with $\lim_{n \to \infty} F^n(x) = u$. Moreover the continuity of F yields

$$u = \lim_{n \to \infty} F^{n+1}(x) = \lim_{n \to \infty} F(F^n(x)) = F(u),$$

therefore u is a fixed point of F. Finally letting $m \to \infty$ in (1.2) yields

$$d(F^n(x), u) \le \frac{L^n}{1 - L}\, d(x, F(x)). \qquad \square$$

Remark 1.1 Theorem 1.1 requires that $L < 1$. If $L = 1$ then F need not have a fixed point as the example $F(x) = x + 1$ for $x \in \mathbf{R}$ shows. We will discuss the case when $L = 1$ in more detail in Chapter 2.

Another natural attempt to extend Theorem 1.1 would be to suppose that $d(F(x), F(y)) < d(x, y)$ for $x, y \in X$ with $x \ne y$. Again F need not have a fixed point as the example $F(x) = \ln(1 + e^x)$ for $x \in \mathbf{R}$ shows. However there is a positive result along these lines in the following theorem of Edelstein.

Theorem 1.2 *Let (X, d) be a compact metric space with $F : X \to X$ satisfying*

$$d(F(x), F(y)) < d(x, y) \text{ for } x,\ y \in X \text{ and } x \ne y.$$

Then F has a unique fixed point in X.

Proof The uniqueness part is easy. To show existence, notice the map $x \mapsto d(x, F(x))$ attains its minimum, say at $x_0 \in X$. We have $x_0 = F(x_0)$ since otherwise

$$d(F(F(x_0)), F(x_0)) < d(F(x_0), x_0)$$

– a contradiction. □

We next present a local version of Banach's contraction principle. This result will be needed in Chapter 3.

Theorem 1.3 *Let (X, d) be a complete metric space and let*

$$B(x_0, r) = \{x \in X : d(x, x_0) < r\}, \text{ where } x_0 \in X \text{ and } r > 0.$$

Suppose $F : B(x_0, r) \to X$ is a contraction (that is, $d(F(x), F(y)) \leq L\, d(x, y)$ for all x, $y \in B(x_0, r)$ with $0 \leq L < 1$) with

$$d(F(x_0), x_0) < (1 - L)\, r.$$

Then F has a unique fixed point in $B(x_0, r)$.

Proof There exists r_0 with $0 \leq r_0 < r$ with $d(F(x_0), x_0) \leq (1 - L)r_0$. We will show that $F : \overline{B(x_0, r_0)} \to \overline{B(x_0, r_0)}$. To see this note that if $x \in \overline{B(x_0, r_0)}$ then

$$\begin{aligned} d(F(x), x_0) &\leq d(F(x), F(x_0)) + d(F(x_0), x_0) \\ &\leq L\, d(x, x_0) + (1 - L)r_0 \leq r_0. \end{aligned}$$

We can now apply Theorem 1.1 to deduce that F has a unique fixed point in $\overline{B(x_0, r_0)} \subset B(x_0, r)$. Again it is easy to see that F has only one fixed point in $B(x_0, r)$. □

Next we examine briefly the behaviour of a contractive map defined on $\overline{B}_r = \overline{B(0, r)}$ (the closed ball of radius r with centre 0) with values in a Banach space E. More general results will be presented in Chapter 3.

Theorem 1.4 *Let $\overline{B}_r$ be the closed ball of radius $r > 0$, centred at zero, in a Banach space E with $F : \overline{B}_r \to E$ a contraction and $F(\partial \overline{B}_r) \subseteq \overline{B}_r$. Then F has a unique fixed point in $\overline{B}_r$.*

Proof Consider

$$G(x) = \frac{x + F(x)}{2}.$$

We first show that $G : \overline{B}_r \to \overline{B}_r$. To see this let

$$x^\star = r\frac{x}{\|x\|} \text{ where } x \in \overline{B}_r \text{ and } x \neq 0.$$

Now if $x \in \overline{B}_r$ and $x \neq 0$,

$$\|F(x) - F(x^\star)\| \leq L\,\|x - x^\star\| = L\,(r - \|x\|),$$

since $x - x^\star = \dfrac{x}{\|x\|}(\|x\| - r)$, and as a result

$$\begin{aligned} \|F(x)\| &\leq \|F(x^\star)\| + \|F(x) - F(x^\star)\| \\ &\leq r + L(r - \|x\|) \leq 2r - \|x\|. \end{aligned}$$

Then for $x \in \overline{B}_r$ and $x \neq 0$

$$\|G(x)\| = \left\| \frac{x + F(x)}{2} \right\| \leq \frac{\|x\| + \|F(x)\|}{2} \leq r.$$

In fact by continuity we also have

$$\|G(0)\| \leq r,$$

and consequently $G : \overline{B}_r \to \overline{B}_r$. Moreover $G : \overline{B}_r \to \overline{B}_r$ is a contraction since

$$\|G(x) - G(y)\| \leq \frac{\|x - y\| + L\|x - y\|}{2} = \frac{[1 + L]}{2}\|x - y\|.$$

Theorem 1.1 implies that G has a unique fixed point $u \in \overline{B}_r$. Of course if $u = G(u)$ then $u = F(u)$. □

Over the last fifty years or so, many authors have given generalisations of Banach's contraction principle. Here for completeness we give one such result. Its proof relies on the following technical result.

Theorem 1.5 *Let (X, d) be a complete metric space and $F : X \to X$ a map (not necessarily continuous). Suppose the following condition holds:*

$$(1.3) \qquad \begin{cases} \text{for each } \epsilon > 0 \text{ there is a } \delta(\epsilon) > 0 \text{ such that if} \\ d(x, F(x)) < \delta(\epsilon), \text{ then } F(B(x, \epsilon)) \subseteq B(x, \epsilon); \\ \text{here } B(x, \epsilon) = \{y \in X : d(x, y) < \epsilon\}. \end{cases}$$

If for some $u \in X$ we have

$$\lim_{n \to \infty} d(F^n(u), F^{n+1}(u)) = 0,$$

then the sequence $\{F^n(u)\}$ converges to a fixed point of F.

Proof Let u be as described above and let $u_n = F^n(u)$. We claim that $\{u_n\}$ is a Cauchy sequence.

Let $\epsilon > 0$ be given. Choose $\delta(\epsilon)$ as in (1.3). We can choose N large enough so that

$$d(u_n, u_{n+1}) < \delta(\epsilon) \text{ for all } n \geq N.$$

Now since $d(u_N, F(u_N)) < \delta(\epsilon)$, then (1.3) guarantees that

$$F(B(u_N, \epsilon)) \subseteq B(u_N, \epsilon),$$

and so $F(u_N) = u_{N+1} \in B(u_N, \epsilon)$. Now by induction

$$F^k(u_N) = u_{N+k} \in B(u_N, \epsilon) \text{ for all } k \in \{0, 1, 2, \ldots\}.$$

Thus

$$d(u_k, u_l) \leq d(u_k, u_N) + d(u_N, u_l) < 2\epsilon \text{ for all } k, l \geq N,$$

and therefore $\{u_n\}$ is a Cauchy sequence. In addition there exists $y \in X$ with $\lim_{n\to\infty} u_n = y$.

We now claim that y is a fixed point of F. Suppose it is not. Then

$$d(y, F(y)) = \gamma > 0.$$

We can now choose (and fix) a $u_n \in B(y, \gamma/3)$ with

$$d(u_n, u_{n+1}) < \delta(\gamma/3).$$

Now (1.3) guarantees that

$$F(B(u_n, \gamma/3)) \subseteq B(u_n, \gamma/3),$$

and consequently $F(y) \in B(u_n, \gamma/3)$. This is a contradiction since

$$d(F(y), u_n) \geq d(F(y), y) - d(u_n, y) > \gamma - \frac{\gamma}{3} = \frac{2\gamma}{3}.$$

Thus $d(y, F(y)) = 0$. □

Theorem 1.6 *Let (X, d) be a complete metric space and let*

$$d(F(x), F(y)) \leq \phi(d(x, y)) \textit{ for all } x, y \in X;$$

here $\phi : [0, \infty) \to [0, \infty)$ is any monotonic, nondecreasing (not necessarily continuous) function with $\lim_{n\to\infty} \phi^n(t) = 0$ for any fixed $t > 0$. Then F has a unique fixed point $u \in X$ with

$$\lim_{n\to\infty} F^n(x) = u \textit{ for each } x \in X.$$

Proof Suppose $t \leq \phi(t)$ for some $t > 0$. Then $\phi(t) \leq \phi(\phi(t))$ and therefore $t \leq \phi^2(t)$. By induction, $t \leq \phi^n(t)$ for $n \in \{1, 2, \ldots\}$. This is a contradiction. Thus $\phi(t) < t$ for each $t > 0$.

In addition,

$$d(F^n(x), F^{n+1}(x)) \leq \phi^n(d(x, F(x))) \text{ for } x \in X,$$

and therefore

$$\lim_{n \to \infty} d(F^n(x), F^{n+1}(x)) = 0 \text{ for each } x \in X.$$

Let $\epsilon > 0$ and choose $\delta(\epsilon) = \epsilon - \phi(\epsilon)$. If $d(x, F(x)) < \delta(\epsilon)$, then for any $z \in B(x, \epsilon) = \{y \in X : d(x, y) < \epsilon\}$ we have

$$\begin{aligned} d(F(z), x) &\leq d(F(z), F(x)) + d(F(x), x) \leq \phi(d(z, x)) + d(F(x), x) \\ &< \phi(d(z, x)) + \delta(\epsilon) \leq \phi(\epsilon) + (\epsilon - \phi(\epsilon)) = \epsilon, \end{aligned}$$

and therefore $F(z) \in B(x, \epsilon)$. Theorem 1.5 guarantees that F has a fixed point u with $\lim_{n \to \infty} F^n(x) = u$ for each $x \in X$. Finally it is easy to see that F has only one fixed point in X. □

Remark 1.2 Note that Theorem 1.1 follows as a special case of Theorem 1.6 if we choose $\phi(t) = Lt$ with $0 \leq L < 1$.

It is natural to begin our applications of fixed point methods with existence and uniqueness of solutions of certain first order initial value problems. In particular we seek solutions to

$$\begin{cases} y'(t) = f(t, y(t)), \\ y(0) = y_0, \end{cases} \tag{1.4}$$

where $f : I \times \mathbf{R}^n \to \mathbf{R}^n$ and $I = [0, b]$. Notice that (1.4) is a system of first order equations because f takes values in $\mathbf{R}^n$.

We begin our analysis of (1.4) by assuming that $f : I \times \mathbf{R}^n \to \mathbf{R}^n$ is continuous. Then, evidently, $y \in C^1(I)$ (the Banach space of functions u whose first derivative is continuous on I and equipped with the norm $|u|_1 = \max\{\sup_{t \in I} |u(t)|, \sup_{t \in I} |u'(t)|\}$) solves (1.4) if and only if $y \in C(I)$ (the Banach space of functions u, continuous on I and equipped with the norm $|u|_0 = \sup_{t \in I} |u(t)|$) solves

$$y(t) = y_0 + \int_0^t f(s, y(s))\, ds. \tag{1.5}$$

Define an integral operator $T : C(I) \to C(I)$ by

$$Ty(t) = y_0 + \int_0^t f(s, y(s))\, ds.$$

Then the equivalence above is expressed briefly by

$$y \text{ solves (1.4) if and only if } y = Ty, \quad T : C(I) \to C(I).$$

In other words, classical solutions to (1.4) are fixed points of the integral operator T. We now present a result known as the Picard–Lindelöf theorem.

Theorem 1.7 *Let $f : I \times \mathbf{R}^n \to \mathbf{R}^n$ be continuous and Lipschitz in y; that is, there exists $\alpha \geq 0$ such that*

$$|f(t,y) - f(t,z)| \leq \alpha\, |y - z| \text{ for all } y,\, z \in \mathbf{R}^n.$$

Then there exists a unique $y \in C^1(I)$ that solves (1.4).

Proof We will apply Theorem 1.1 to show that T has a unique fixed point. At first glance it seems natural to use the maximum norm on $C(I)$, but this choice would lead us only to a local solution defined on a subinterval of I. The trick is to use the weighted maximum norm

$$\|y\|_\alpha = |e^{-\alpha t} y(t)|_0$$

on $C(I)$. Observe that $C(I)$ is a Banach space with this norm since it is equivalent to the maximum norm, that is,

$$e^{-\alpha b} |y|_0 \leq \|y\|_\alpha \leq |y|_0.$$

We now show that T is a contraction on $(C(I), \|\cdot\|_\alpha)$. To see this let $y, z \in C(I)$ and notice

$$Ty(t) - Tz(t) = \int_0^t [f(s, y(s)) - f(s, z(s))]\, ds \text{ for } t \in I.$$

Thus for $t \in I$,

$$\begin{aligned}
e^{-\alpha t} |(Ty - Tz)(t)| &\leq e^{-\alpha t} \int_0^t \alpha e^{\alpha s} e^{-\alpha s} |y(s) - z(s)|\, ds \\
&\leq e^{-\alpha t} \left(\int_0^t \alpha e^{\alpha s}\, ds \right) \|y - z\|_\alpha \\
&\leq e^{-\alpha t} \left(e^{\alpha t} - 1 \right) \|y - z\|_\alpha \\
&\leq (1 - e^{-\alpha b}) \|y - z\|_\alpha,
\end{aligned}$$

and therefore

$$\|Ty - Tz\|_\alpha \leq \left(1 - e^{-\alpha b}\right) \|y - z\|_\alpha.$$

Since $1 - e^{-\alpha b} < 1$, the Banach contraction principle implies that there is a unique $y \in C(I)$ with $y = Ty$; equivalently (1.4) has a unique solution $y \in C^1(I)$. □

Now we relax the continuity assumption on f and extend the notion of a solution of (1.4) accordingly. We want to do this in a way that preserves the natural equivalence between (1.4) and the equation $y = Ty$, which was obtained by integrating. To this end we follow the ideas of Carathéodory and make the following definitions.

Definition 1.1 A function $y \in W^{1,p}(I)$ is an *L^p-Carathéodory solution* of (1.4) if y solves (1.4) in the almost everywhere sense on I; here $W^{1,p}(I)$ is the Sobolev class of functions u, with u absolutely continuous and $u' \in L^p(I)$.

Definition 1.2 A function $f : I \times \mathbf{R}^n \to \mathbf{R}^n$ is an *L^p-Carathéodory function* if it satisfies the following conditions:

(c1) the map $y \mapsto f(t, y)$ is continuous for almost every $t \in I$;
(c2) the map $t \mapsto f(t, y)$ is measurable for all $y \in \mathbf{R}^n$;
(c3) for every $c > 0$ there exists $h_c \in L^p(I)$ such that $|y| \leq c$ implies that $|f(t, y)| \leq h_c(t)$ for almost every $t \in I$.

If f is an L^p-Carathéodory function, then $y \in W^{1,p}(I)$ solves (1.4) if and only if

$$y \in C(I) \text{ and } y(t) = y_0 + \int_0^t f(s, y(s))\, ds.$$

In fact (c1) and (c2) imply that the integrand on the right is measurable for any measurable y, and (c3) guarantees that it is integrable for any bounded measurable y. The stated equivalence now is clear. Therefore just as in the continuous case,

$$\text{(1.4) has a solution } y \text{ if and only if } y = Ty, \;\; T : C(I) \to C(I).$$

Theorem 1.8 *Let $f : I \times \mathbf{R}^n \to \mathbf{R}^n$ be an L^p-Carathéodory function and L^p-Lipschitz in y; that is, there exists $\alpha \in L^p(I)$ with*

$$|f(t, y) - f(t, z)| \leq \alpha(t)|y - z| \text{ for all } y,\ z \in \mathbf{R}^n.$$

Then there exists a unique $y \in W^{1,p}(I)$ that solves (1.4).

Proof The proof is similar to Theorem 1.7 and will only be sketched here. Let

$$A(t) = \int_0^t \alpha(s)\, ds.$$

Then $A'(t) = \alpha(t)$ for a.e. t. Define

$$\|y\|_A = \left|e^{-A(t)}y(t)\right|_0.$$

The norm is equivalent to the maximum norm because

$$e^{-\|\alpha\|_1}|y|_0 \le \|y\|_A \le |y|_0, \text{ where } \|\alpha\|_1 = \int_0^b |\alpha(t)|\, dt.$$

Thus $(C(I), \|\cdot\|_A)$ is a Banach space and use of the Banach contraction principle, essentially as in the proof of Theorem 1.7, implies that there exists a unique $y \in C(I)$ with $y = Ty$. It follows that (1.4) has a unique L^p-Carathéodory solution on I. □

Notes Most of the results in Chapter 1 may be found in the classical books of Dugundji and Granas [55], Goebel and Kirk [77] and Zeidler [191].

Exercises

1.1 Show that a contraction F from an incomplete metric space into itself need not have a fixed point.

1.2 Let (X, d) be a complete metric space and let $F : X \to X$ be such that $F^N : X \to X$ is a contraction for some positive integer N. Show that F has a unique fixed point $u \in X$ and that for each $x \in X$, $\lim_{n\to\infty} F^n(x) = u$.

1.3 Using the result obtained in Exercise 1.2, give an alternative proof for the Picard–Lindelöf theorem (Theorem 1.7).

1.4 Let $\overline{B}_r$ be the closed ball of radius $r > 0$, centred at zero, in a Banach space E with $F : \overline{B}_r \to E$ a contraction and $F(-x) = -F(x)$ for $x \in \partial\overline{B}_r$. Show F has a fixed point in $\overline{B}_r$.

1.5 Let U be an open subset of a Banach space E and let $F : U \to E$ be a contraction. Show that $(I - F)(U)$ is open.

1.6 Let (X, d) be a complete metric space, P a topological space and $F : X \times P \to X$. Suppose F is a contraction uniformly over P (that is, for each $x, y \in X$, $d(F(x,p), F(y,p)) \leq L\, d(x,y)$ for all $p \in P$) and is continuous in p for each fixed $x \in X$. Let x_p be the unique fixed point of $F_p : X \to X$, where $F_p(x) = F(x,p)$. Show that $p \mapsto x_p$ is continuous.

1.7 Let $k : [0,1] \times [0,1] \times \mathbf{R} \to \mathbf{R}$ be continuous with

$$|k(t,s,x) - k(t,s,y)| \leq L\,|x - y|$$

for all $(t,s) \in [0,1] \times [0,1]$ and $x, y \in \mathbf{R}$ (here $L \geq 0$ is a constant) and $v \in C[0,1]$.

(a) Show that

$$u(t) = v(t) + \int_0^t k(t,s,u(s))\,ds, \quad 0 \leq t \leq 1,$$

has a unique solution $u \in C[0,1]$.

(b) Choose $u_0 \in C[0,1]$ and define a sequence of functions $\{u_n\}$ inductively by

$$u_{n+1}(t) = v(t) + \int_0^t k(t,s,u_n(s))\,ds, \quad n = 0, 1, \ldots.$$

Show that the sequence $\{u_n\}$ converges uniformly on $[0,1]$ to the unique solution $u \in C[0,1]$.

1.8 Let (X, d) be a complete metric space and let $\phi : X \to [0, \infty)$ be a map (not necessarily continuous). Suppose

$$\inf\{\phi(x) + \phi(y) : d(x,y) \geq \gamma\} = \mu(\gamma) > 0 \text{ for all } \gamma > 0.$$

Show that each sequence $\{x_n\}$ in X, for which $\lim_{n\to\infty} \phi(x_n) = 0$, converges to one and only one point $u \in X$.

1.9 Let (X, d) be a complete metric space and let $F : X \to X$ be continuous. Suppose $\phi(x) = d(x, F(x))$ satisfies

$$\inf\{\phi(x) + \phi(y) : d(x,y) \geq \gamma\} = \mu(\gamma) > 0 \text{ for all } \gamma > 0,$$

and that $\inf_{x\in X} d(x, F(x)) = 0$. Show that F has a unique fixed point.

1.10 If in Theorem 1.6 the assumptions on ϕ are replaced by $\phi : [0,\infty) \to [0,\infty)$ is upper semicontinuous from the right on $[0,\infty)$ (that is, $\limsup_{s\to t^+} \phi(s) \leq \phi(t)$ for $t \in [0,\infty)$) and satisfies

$\phi(t) < t$ for $t > 0$. Show that F has a unique fixed point $u \in X$ with $\lim_{n\to\infty} F^n(x) = u$ for each $x \in X$.

1.11 Let T be a map of the metric space (X, ρ) into itself such that, for a fixed positive integer n,

$$\rho(T^n x, T^n y) \leq \alpha^n \rho(x, y) \text{ for } x, y \in X;$$

here α is a positive real number. Show that the function σ defined by

$$\sigma(x, y) := \rho(x, y) + \frac{1}{\alpha}\rho(Tx, Ty) + \cdots + \frac{1}{\alpha^{n-1}}\rho(T^{n-1}x, T^{n-1}y)$$

is a metric on X and T satisfies

$$\sigma(Tx, Ty) \leq \alpha\sigma(x, y) \text{ for } x, y \in X.$$

2
Nonexpansive Maps

Let (X, d) be a metric space with $C \subseteq X$. Recall that a mapping $F : C \to X$ is nonexpansive if $L = 1$, that is if F satisfies

$$d(F(x), F(y)) \leq d(x, y) \text{ for all } x, y \in X.$$

In Chapter 1 we briefly alluded to nonexpansive maps and gave an example of a nonexpansive map free of fixed points.

We begin this chapter by presenting a result known as Schauder's theorem for nonexpansive maps. It is a special case of Schauder's fixed point theorem which will be presented in Chapter 4.

Theorem 2.1 *Let C be a nonempty, closed, convex subset of a normed linear space E with $F : C \to C$ nonexpansive and $F(C)$ a subset of a compact set of C. Then F has a fixed point.*

Proof Let $x_0 \in C$. For $n = 2, 3, \ldots$, define

$$F_n := \left(1 - \frac{1}{n}\right) F + \frac{1}{n} x_0.$$

Since C is convex and $x_0 \in C$, we see that $F_n : C \to C$ and it is clear that $F_n : C \to C$ is a contraction. Therefore by Theorem 1.1 each F_n has a unique fixed point $x_n \in C$, that is,

$$x_n = F_n(x_n) = \left(1 - \frac{1}{n}\right) F(x_n) + \frac{1}{n} x_0.$$

In addition, since $F(C)$ lies in a compact subset of C, there exist a subsequence S of integers and a $u \in C$ with

$$F(x_n) \to u \text{ as } n \to \infty \text{ in } S.$$

Thus

$$x_n = \left(1 - \frac{1}{n}\right) F(x_n) + \frac{1}{n} x_0 \to u \text{ as } n \to \infty \text{ in } S.$$

By continuity

$$F(x_n) \to F(u) \text{ as } n \to \infty \text{ in } S,$$

and therefore $u = F(u)$. □

The main theorem of this section is a result proved independently by Browder, Göhde and Kirk in 1965. We state it as follows.

Theorem 2.2 *Let C be a nonempty, closed, bounded, convex set in a (real) Hilbert space H. Then each nonexpansive map $F : C \to C$ has at least one fixed point.*

Remark 2.1 Notice that uniqueness need not hold as the example $F(x) = x$, $x \in C = [0, 1]$, shows.

Remark 2.2 In fact in Theorem 2.2 it is enough to assume that H is a uniformly convex Banach space (see Exercise 2.9).

In the proof of Theorem 2.2 we will need the following two technical results.

Theorem 2.3 *Let H be a Hilbert space with u, $v \in H$, and let r, R be constants with $0 \le r \le R$. If there exists an $x \in H$ with*

$$\|u - x\| \le R, \ \|v - x\| \le R \text{ and } \left\|\frac{u+v}{2} - x\right\| \ge r,$$

then

$$\|u - v\| \le 2\sqrt{R^2 - r^2}.$$

Proof The parallelogram law gives

$$\begin{aligned} \|u - v\|^2 &= 2\|u - x\|^2 + 2\|v - x\|^2 - \|(u - x) + (v - x)\|^2 \\ &\le 2R^2 + 2R^2 - 4\left\|\frac{u+v}{2} - x\right\|^2 \le 4(R^2 - r^2). \end{aligned}$$

□

Theorem 2.4 *Let H be a Hilbert space, $C \subseteq H$ a bounded set and $F : C \to C$ a nonexpansive map. Suppose $x \in C$, $y \in C$ and $a = \dfrac{x+y}{2} \in C$.*

Let $\delta(C)$ denote the diameter of C and let $\epsilon \le \delta(C)$ with $\|x - F(x)\| \le \epsilon$ and $\|y - F(y)\| \le \epsilon$. Then

$$\|a - F(a)\| \le 2\sqrt{\epsilon}\sqrt{2\delta(C)}.$$

Proof Since

$$\|x - y\| \le \left\| x - \frac{a + F(a)}{2} \right\| + \left\| y - \frac{a + F(a)}{2} \right\|,$$

we may assume without loss of generality that

$$\left\| x - \frac{a + F(a)}{2} \right\| \ge \frac{1}{2}\|x - y\|.$$

However since

$$\|a - x\| = \frac{1}{2}\|x - y\|,$$

we have

$$\begin{aligned} \|F(a) - x\| &\le \|F(a) - F(x)\| + \|F(x) - x\| \\ &\le \|a - x\| + \epsilon = \frac{1}{2}\|x - y\| + \epsilon. \end{aligned}$$

Theorem 2.3 with $r = \frac{1}{2}\|x - y\|$, $R = \frac{1}{2}\|x - y\| + \epsilon$, $u = a$ and $v = F(a)$ gives

$$\begin{aligned} \|a - F(a)\| &\le 2\sqrt{\left(\frac{1}{2}\|x - y\| + \epsilon\right)^2 - \left(\frac{1}{2}\|x - y\|\right)^2} \\ &= 2\sqrt{\|x - y\|\epsilon + \epsilon^2} = 2\sqrt{\epsilon}\sqrt{\|x - y\| + \epsilon} \\ &\le 2\sqrt{\epsilon}\sqrt{2\delta(C)}. \end{aligned}$$

□

Proof of Theorem 2.2 Assume that $0 \in C$. (A modified argument from the one given below holds for any $x_0 \in C$, therefore for simplicity we let $x_0 = 0$.) Also assume that $F(0) \neq 0$ (otherwise we are finished). For each $n = 2, 3, \ldots$, notice that

$$F_n := \left(1 - \frac{1}{n}\right) F : C \to C$$

is a contraction. Now Theorem 1.1 guarantees that there exists a unique $x_n \in C$ with

$$x_n = F_n(x_n) = \left(1 - \frac{1}{n}\right) F(x_n).$$

Thus

$$\|x_n - F(x_n)\| = \frac{1}{n}\,\|F(x_n)\| \le \frac{1}{n}\delta(C), \tag{2.1}$$

where $\delta(C)$ denotes the diameter of C. For each $n \in \{2, 3, \ldots\}$, let

$$Q_n = \left\{x \in C : \|x - F(x)\| \le \frac{1}{n}\delta(C)\right\}.$$

Now

$$Q_2 \supseteq Q_3 \supseteq \cdots \supseteq Q_n \supseteq \cdots$$

is a decreasing sequence of nonempty (see (2.1)) closed sets. Let

$$d_n = \inf\{\|x\| : x \in Q_n\}$$

and since the Q_ns are decreasing we have

$$d_2 \le d_3 \le \cdots \le d_n \le \cdots, \text{ with } d_i \le \delta(C)$$

for each $i \in \{2, 3, \ldots\}$. Consequently, $d_n \uparrow d$ with $d \le \delta(C)$. Next let

$$A_n = Q_{8n^2} \cap \overline{B(0, d + 1/n)},$$

where

$$B(0, d + 1/n) = \left\{x \in H : \|x\| < d + \frac{1}{n}\right\}.$$

Now A_n is a decreasing sequence of closed, nonempty sets. We now show that $\lim_{n\to\infty} \delta(A_n) = 0$. To see this let $u, v \in A_n$. Then

$$\|u - 0\| \le d + \frac{1}{n} \text{ and } \|v - 0\| \le d + \frac{1}{n}. \tag{2.2}$$

Also since $u, v \in Q_{8n^2}$ we have

$$\|u - F(u)\| \le \frac{1}{8n^2}\delta(C) \text{ and } \|v - F(v)\| \le \frac{1}{8n^2}\delta(C).$$

Thus Theorem 2.4 implies that

$$\left\|\frac{u+v}{2} - F\left(\frac{u+v}{2}\right)\right\| \le 2\sqrt{2\delta(C)}\sqrt{\frac{1}{8n^2}\delta(C)} = \frac{1}{n}\delta(C),$$

therefore $\frac{u+v}{2} \in Q_n$ and

$$\left\|\frac{u+v}{2} - 0\right\| \ge d_n. \tag{2.3}$$

Now (2.2), (2.3) and Theorem 2.3 imply

$$\|u - v\| \le 2\sqrt{\left(d + \frac{1}{n}\right)^2 - {d_n}^2},$$

and therefore

$$\delta(A_n) \le 2\sqrt{\frac{2d}{n} + \frac{1}{n^2} + (d^2 - {d_n}^2)}.$$

Consequently $\lim_{n\to\infty} \delta(A_n) = 0$. Cantor's theorem (applied to $\{A_n\}_{n=2}^{\infty}$) guarantees the existence of an

$$x_0 \in \bigcap_{n=2}^{\infty} A_n.$$

Since

$$x_0 \in \bigcap_{n=2}^{\infty} Q_{8n^2}$$

we have

$$\|x_0 - F(x_0)\| \le \frac{\delta(C)}{8n^2} \text{ for all } n \in \{2, 3, \ldots\}.$$

Therefore

$$\|x - F(x_0)\| = 0$$

and the theorem is proved. □

We now examine the behaviour of nonexpansive maps defined on $\overline{B}_r$ (the closed ball of radius r and centre 0) with values in the Hilbert space H. In particular we prove a nonlinear alternative of Leray–Schauder type for nonexpansive maps. It is worth remarking that a more general result will be presented in Chapter 3.

Theorem 2.5 *Let H be a real Hilbert space and $\overline{B}_r = \{x \in H : \|x\| \le r\}$ with $r > 0$. Then each nonexpansive map $F : \overline{B}_r \to H$ has at least one of the following two properties:*

(A1) *F has a fixed point in $\overline{B}_r$,*

(A2) *There are an $x \in \partial\overline{B}_r$ and a $\lambda \in (0, 1)$ with $x = \lambda F(x)$.*

Proof Define a map (radial retraction) $r : H \to \overline{B}_r$ by

$$r(x) = \begin{cases} x, & \|x\| \le r, \\ r\dfrac{x}{\|x\|}, & \|x\| > r. \end{cases}$$

It is easy to check (see Exercise 2.1) that $r : H \to \overline{B}_r$ is nonexpansive. As a result $r \circ F : \overline{B}_r \to \overline{B}_r$ is a nonexpansive map. Theorem 2.2 guarantees that there exists $x \in \overline{B}_r$ with $r(F(x)) = x$. If $F(x) \in \overline{B}_r$, then

$$x = r(F(x)) = F(x),$$

and F has a fixed point, that is, (A1) occurs. If $F(x)$ does not belong to $\overline{B}_r$ then

$$x = r(F(x)) = r\frac{F(x)}{\|F(x)\|} = \lambda F(x) \text{ with } \lambda = \frac{r}{\|F(x)\|} < 1,$$

that is, (A2) occurs since $x \in \partial\overline{B}_r$. □

It is easy to put conditions on F to guarantee that the second possibility (A2) in Theorem 2.5 does not occur.

Theorem 2.6 *Let H be a real Hilbert space, $\overline{B}_r = \{x \in H : \|x\| \leq r\}$ with $r > 0$, and let $F : \overline{B}_r \to H$ be nonexpansive. Suppose for all $x \in \partial\overline{B}_r$ one of the following four conditions holds:*

(i) $\|F(x)\| \leq \|x\|$,
(ii) $\|F(x)\| \leq \|x - F(x)\|$,
(iii) $\|F(x)\|^2 \leq \|x\|^2 + \|x - F(x)\|^2$,
(iv) $\langle x, F(x)\rangle \leq \|x\|^2$.

Then F has a fixed point in $\overline{B}_r$.

Proof We prove the result with (ii) holding. If F has no fixed points then by Theorem 2.5 there exist $z \in \partial\overline{B}_r$ and $\lambda \in (0,1)$ with $z = \lambda F(z)$. In particular $F(z) \neq 0$ and

$$\|F(z)\| = \|F(\lambda F(z))\| \leq \|\lambda F(z) - F(\lambda F(z))\|,$$

that is,

$$\|F(z)\| \leq (1 - \lambda)\|F(z)\|,$$

therefore $1 \leq 1 - \lambda$. This is a contradiction. □

Notes The approach in Chapter 2 has been adapted from Dugundji and Granas [55]. Other relevant approaches may be found in Goebel and Kirk [77] and Smart [174].

Exercises

2.1 Show that the radial retraction defined in Theorem 2.5 is nonexpansive.

2.2 Complete the proof of Theorem 2.6.

2.3 Let (X, d) be a metric space and let $F : K \to Q$ be a continuous mapping; here K is a closed subset of X and Q is a compact subset of X. Suppose for each $\epsilon > 0$ there exists an $x(\epsilon) \in K$ with $d(F(x(\epsilon)), x(\epsilon)) < \epsilon$. Show that F has a fixed point.

2.4 Let C be a closed, bounded, convex subset of a Banach space E and $F : C \to C$ a nonexpansive map. Show that for each $\epsilon > 0$, there exists $x(\epsilon) \in C$ with $\|F(x(\epsilon)) - x(\epsilon)\| < \epsilon$.

2.5 Let H be a Hilbert space and $C \subseteq H$ a closed, convex set. Define $P_C : H \to C$ to be the map sending each $x \in H$ to the nearest point in C. Show that $P_C : H \to C$ is nonexpansive.

2.6 Let K be a bounded, convex subset of a uniformly convex Banach space X and $F : K \to X$ a nonexpansive map. Suppose for $\{u_n\}$, $\{v_n\}$ in K and $z_n = \dfrac{u_n + v_n}{2}$, we have

$$\lim_{n\to\infty} \|u_n - F(u_n)\| = 0 \text{ and } \lim_{n\to\infty} \|v_n - F(v_n)\| = 0.$$

Show that $\lim_{n\to\infty} \|z_n - F(z_n)\| = 0$.

2.7 Let K be a bounded, closed, convex subset of a uniformly convex Banach space X and $F : K \to X$ be a nonexpansive map with

$$\inf\{\|x - F(x)\| : x \in K\} = 0.$$

Show that F has a fixed point in K.

2.8 Let X be a uniformly convex Banach space and K a closed, convex subset of X with $F : K \to X$ a nonexpansive map. Suppose that $\{u_j\}$ is a weakly convergent sequence in K with weak limit u_0. If $(I - F)(u_j)$ converges strongly to an element w_0 in K, then show that $(I - F)(u_0) = w_0$. (This means that $I - F$ is demiclosed on K.)

2.9 Let X be a uniformly convex Banach space and K a closed, convex, bounded subset of X. Prove that every nonexpansive map $F : K \to K$ has a fixed point.

3

Continuation Methods for Contractive and Nonexpansive Mappings

We begin this chapter by showing that the property of having a fixed point is invariant by homotopy for contractions.

Let (X, d) be a complete metric space and U an open subset of X.

Definition 3.1 Let $F : \overline{U} \to X$ and $G : \overline{U} \to X$ be two contractions; here $\overline{U}$ denotes the closure of U in X. We say that F and G are *homotopic* if there exists $H : \overline{U} \times [0,1] \to X$ with the following properties:

(a) $H(\cdot, 0) = G$ and $H(\cdot, 1) = F$;
(b) $x \neq H(x,t)$ for every $x \in \partial U$ and $t \in [0,1]$ (here ∂U denotes the boundary of U in X);
(c) there exists α, $0 \le \alpha < 1$, such that $d(H(x,t), H(y,t)) \le \alpha\, d(x,y)$ for every $x, y \in \overline{U}$ and $t \in [0,1]$;
(d) there exists M, $M \ge 0$, such that $d(H(x,t), H(x,s)) \le M|t-s|$ for every $x \in \overline{U}$ and $t, s \in [0,1]$.

Theorem 3.1 *Let (X, d) be a complete metric space and U an open subset of X. Suppose that $F : \overline{U} \to X$ and $G : \overline{U} \to X$ are two homotopic contractive maps and G has a fixed point in U. Then F has a fixed point in U.*

Proof Consider the set

$$A = \{\lambda \in [0,1] : x = H(x, \lambda) \text{ for some } x \in U\}$$

where H is a homotopy between F and G as described in Definition 3.1. Notice A is nonempty since G has a fixed point, that is, $0 \in A$. We will show that A is both open and closed in $[0,1]$ and hence by connectedness we have that $A = [0,1]$. As a result, F has a fixed point in U.

We first show that A is closed in $[0,1]$. To see this let

$$\{\lambda_n\}_{n=1}^{\infty} \subseteq A \text{ with } \lambda_n \to \lambda \in [0,1] \text{ as } n \to \infty.$$

We must show that $\lambda \in A$. Since $\lambda_n \in A$ for $n = 1, 2, \ldots$, there exists $x_n \in U$ with $x_n = H(x_n, \lambda_n)$. Also for $n, m \in \{1, 2, \ldots\}$ we have

$$\begin{aligned} d(x_n, x_m) &= d(H(x_n, \lambda_n), H(x_m, \lambda_m)) \\ &\leq d(H(x_n, \lambda_n), H(x_n, \lambda_m)) + d(H(x_n, \lambda_m), H(x_m, \lambda_m)) \\ &\leq M|\lambda_n - \lambda_m| + \alpha\, d(x_n, x_m), \end{aligned}$$

that is,

$$d(x_n, x_m) \leq \left(\frac{M}{1-\alpha}\right) |\lambda_n - \lambda_m|.$$

Since $\{\lambda_n\}$ is a Cauchy sequence we have that $\{x_n\}$ is also a Cauchy sequence, and since X is complete there exists $x \in \overline{U}$ with $\lim_{n\to\infty} x_n = x$. In addition, $x = H(x, \lambda)$ since

$$\begin{aligned} d(x_n, H(x, \lambda)) &= d(H(x_n, \lambda_n), H(x, \lambda)) \\ &\leq M|\lambda_n - \lambda| + \alpha\, d(x_n, x). \end{aligned}$$

Thus $\lambda \in A$ and A is closed in $[0,1]$.

Next we show that A is open in $[0,1]$. Let $\lambda_0 \in A$. Then there exists $x_0 \in U$ with $x_0 = H(x_0, \lambda_0)$. Fix $\epsilon > 0$ such that

$$\epsilon \leq \frac{(1-\alpha)r}{M} \text{ where } r < \operatorname{dist}(x_0, \partial U),$$

and where $\operatorname{dist}(x_0, \partial U) = \inf\{d(x_0, x) : x \in \partial U\}$. Fix $\lambda \in (\lambda_0 - \epsilon, \lambda_0 + \epsilon)$. Then for $x \in \overline{B(x_0, r)} = \{x : d(x, x_0) \leq r\}$,

$$\begin{aligned} d(x_0, H(x, \lambda)) &\leq d(H(x_0, \lambda_0), H(x, \lambda_0)) + d(H(x, \lambda_0), H(x, \lambda)) \\ &\leq \alpha\, d(x_0, x) + M|\lambda - \lambda_0| \\ &\leq \alpha r + (1-\alpha)r = r. \end{aligned}$$

Thus for each fixed $\lambda \in (\lambda_0 - \epsilon, \lambda_0 + \epsilon)$,

$$H(\cdot, \lambda) : \overline{B(x_0, r)} \to \overline{B(x_0, r)}.$$

We can now apply Theorem 1.1 (an argument based on Theorem 1.3 could also be used) to deduce that $H(\cdot, \lambda)$ has a fixed point in U. Thus $\lambda \in A$ for any $\lambda \in (\lambda_0 - \epsilon, \lambda_0 + \epsilon)$ and therefore A is open in $[0,1]$. $\square$

For the remainder of this chapter we will assume that X is a Banach space. We now present a nonlinear alternative of Leray–Schauder type for contractive maps.

Theorem 3.2 *Suppose U is an open subset of a Banach space X, $0 \in U$ and $F : \overline{U} \to X$ a contraction with $F(\overline{U})$ bounded. Then either*

(A1) *F has a fixed point in $\overline{U}$, or*
(A2) *there exist $\lambda \in (0,1)$ and $u \in \partial U$ with $u = \lambda F(u)$*

holds.

Proof Assume (A2) does not hold and F has no fixed points on ∂U (otherwise we are finished). Then

$$u \neq \lambda F(u) \text{ for all } u \in \partial U \text{ and } \lambda \in [0,1].$$

Let $H : \overline{U} \times [0,1] \to X$ be given by

$$H(x,t) = tF(x),$$

and let G be the zero map. Notice G has a fixed point in U (that is, $0 = G(0)$) and F and G are two homotopic, contractive mappings. We can now apply Theorem 3.1 to deduce that there exists $x \in U$ with $x = F(x)$, that is, (A1) occurs. □

It is natural to ask whether we can extend Theorem 3.2 to nonexpansive maps as Theorem 2.5 suggests.

Theorem 3.3 *Let U be a bounded, open, convex subset of a uniformly convex Banach space X, with $0 \in U$ and $F : \overline{U} \to X$ a nonexpansive map. Then either*

(A1) *F has a fixed point in $\overline{U}$, or*
(A2) *there exist $\lambda \in (0,1)$ and $u \in \partial U$ with $u = \lambda F(u)$*

is true.

Proof Assume (A2) does not hold. Consider for each $n \in \{2, 3, \ldots\}$, the mapping

$$F_n := \left(1 - \frac{1}{n}\right) F : \overline{U} \to X.$$

Notice that F_n is a contraction with contraction constant $1 - 1/n$. Applying Theorem 3.2 to F_n, we deduce that either F_n has a fixed point in

U, or there exist $\lambda \in (0,1)$ and $u \in \partial U$ with $u = \lambda F_n(u)$. Suppose the latter is true, that is, there exist

$$\lambda \in (0,1) \text{ and } u \in \partial U \text{ with } u = \lambda F_n(u).$$

Then

$$u = \lambda\left(1 - \frac{1}{n}\right) F(u) = \eta F(u) \text{ where } 0 < \eta = \lambda\left(1 - \frac{1}{n}\right) < 1$$

– a contradiction since property (A2) does not occur. Consequently for each $n \in \{2, 3, \ldots\}$ we have that F_n has a fixed point $u_n \in U$. A standard result (if E is a reflexive Banach space, then any norm bounded sequence in E has a weakly convergent subsequence) implies (since $\overline{U}$ is closed, bounded and convex – hence weakly closed) that there exist a subsequence S of integers and a $u \in \overline{U}$ with

$$u_n \rightharpoonup u \text{ as } n \to \infty \text{ in } S; \text{ here } \rightharpoonup \text{ denotes weak convergence.}$$

In addition since $u_n = (1 - 1/n)F(u_n)$ we have

$$\begin{aligned}
\|(I - F)(u_n)\| &= \frac{1}{n}\|F(u_n)\| \\
&\leq \frac{1}{n}(\|F(u_n) - F(0)\| + \|F(0)\|) \\
&\leq \frac{1}{n}(\|u_n\| + \|F(0)\|).
\end{aligned}$$

Thus $(I - F)(u_n)$ converges strongly to 0. The demiclosedness of $I - F$ (see Exercise 2.8) implies that $u = F(u)$, and as a result (A1) occurs. □

To illustrate how Theorem 3.2 can be applied in practice we turn our attention to the second order homogeneous Dirichlet problem,

$$\text{(3.1)} \qquad \begin{cases} y'' = f(t, y, y') \text{ for } t \in [a, b], \\ y(a) = y(b) = 0, \end{cases}$$

where $f : [a, b] \times \mathbf{R}^2 \to \mathbf{R}$ is continuous. Associated with (3.1) we consider the following related family of problems:

$$\text{(3.2)}_\lambda \qquad \begin{cases} y'' = \lambda f(t, y, y') \text{ for } t \in [a, b], \\ y(a) = y(b) = 0, \end{cases}$$

for $\lambda \in (0, 1)$.

Define an operator $F : C^1[a, b] \to C^1[a, b]$ by

$$Fy(t) := \int_a^b G(t, s) f(s, y(s), y'(s))\, ds$$

where the Green's function $G(t,s)$ is given by

$$G(t,s) = \begin{cases} -\dfrac{(t-a)(b-s)}{b-a}, & a \le t \le s \le b, \\ -\dfrac{(s-a)(b-t)}{b-a}, & a \le s \le t \le b. \end{cases}$$

By the properties of the Green's function, the fixed points of F are the classical solutions of (3.1). Under an appropriate local Lipschitz condition on f, we will use the nonlinear alternative for contractive maps to establish that F restricted to the closure of a suitable open set $U \subseteq C^1[a,b]$ is contractive and has a fixed point (in fact a unique fixed point) in $\overline{U}$. Hence (3.1) has a unique solution in $\overline{U}$.

To this end we assume that f satisfies the following local Lipschitz condition:

$$(3.3) \qquad \begin{cases} \text{there are a subset } D \subseteq \mathbf{R}^2 \text{ and constants } K_0 \text{ and } K_1 \\ \text{such that } f \text{ restricted to } [a,b] \times D \text{ satisfies} \\ |f(t,y,y') - f(t,z,z')| \le K_0|y-z| + K_1|y'-z'|. \end{cases}$$

Define a modified maximum norm on $C^1[a,b]$ by

$$\|y\| = K_0|y|_0 + K_1|y'|_0 \text{ where } |y|_0 = \sup_{t\in[a,b]} |y(t)| \text{ and } |y'|_0 = \sup_{t\in[a,b]} |y'(t)|.$$

For functions y and z whose values and derivative values lie in the region where f is locally Lipschitz, we have

$$\begin{aligned} |(Fy - Fz)(t)| &= \left| \int_a^b G(t,s)[f(s,y(s),y'(s)) - f(s,z(s),z'(s))]\,ds \right| \\ &\le \frac{(b-a)^2}{8}\|y-z\|, \end{aligned}$$

since

$$\max_{t\in[a,b]} \int_a^b |G(t,s)|\,ds = \max_{t\in[a,b]} \frac{(b-t)(t-a)}{2} = \frac{(b-a)^2}{8}.$$

Thus

$$|Fy - Fz|_0 \le \frac{(b-a)^2}{8}\|y-z\|$$

for functions y and z whose values and derivative values lie in the region where f is locally Lipschitz. Likewise

$$|(Fy-Fz)'|_0 \le \frac{(b-a)}{2}\|y-z\|$$

for functions y and z whose values and derivative values lie in the region where f is locally Lipschitz, since

$$\max_{t\in[a,b]} \int_a^b |G_t(t,s)|\,ds = \max_{t\in[a,b]} \frac{(b-t)^2+(t-a)^2}{2(b-a)} = \frac{b-a}{2}.$$

Consequently

$$\|Fy - Fz\| \le \left[K_0\frac{(b-a)^2}{8} + K_1\frac{(b-a)}{2}\right]\|y-z\|, \tag{3.4}$$

for functions y and z whose values and derivative values lie in the region where f is locally Lipschitz. This inequality and Theorem 3.2 enable us to establish the following existence and uniqueness principle for (3.1).

Theorem 3.4 *Let $f : [a,b] \times \mathbf{R}^2 \to \mathbf{R}$ be continuous and satisfy* (3.3) *in a set D with constants K_0 and K_1 such that*

$$K_0\frac{(b-a)^2}{8} + K_1\frac{(b-a)}{2} < 1 \tag{3.5}$$

is true. Suppose there is a bounded open set of functions $U \subseteq C^1[a,b]$ with $0 \in U$ such that

$$u \in \overline{U} \textit{ implies } (u(t), u'(t)) \in D \textit{ for all } t \in [a,b] \tag{3.6}$$

and

$$y \textit{ solves } (3.2)_\lambda \textit{ for some } \lambda \in (0,1) \textit{ implies } y \notin \partial U \tag{3.7}$$

hold. Then (3.1) *has a unique solution in $\overline{U}$.*

Proof Evidently $F : \overline{U} \to C^1[a,b]$ is contractive by (3.4) and (3.5). Apply Theorem 3.2 and note that (A2) cannot occur because of (3.7). □

Remark 3.1 In many important applications, the function f is independent of y'; that is $f = f(t,y)$. In this case, a straightforward review of the reasoning given above shows that we can regard F as $F : C[a,b] \to C[a,b]$. This leads to a useful variant of Theorem 3.4 in which $D \subseteq \mathbf{R}$, all reference to y' and z' is dropped in (3.3), and $U \subseteq C[a,b]$.

Example 3.1 The boundary value problem

$$\begin{cases} y''(t) = -e^{y(t)}, & t \in [0,1], \\ y(0) = y(1) = 0 \end{cases} \tag{3.8}$$

has a unique solution with maximum norm *at most* 1. We note that (3.8) models the steady state temperature in a rod with temperature dependent internal heating.

To establish the above claim we apply Theorem 3.4 and Remark 3.1 with $f = f(t, y) = -e^y$. By the mean value theorem we have that

$$|y| \leq 1 \text{ and } |z| \leq 1 \text{ imply } |e^y - e^z| \leq e^{\max\{y,z\}}|y - z| \leq e|y - z|.$$

We take

$$D = [-1, 1] \text{ and } U = \left\{ y \in C[0,1] : |y|_0 = \sup_{t \in [0,1]} |y(t)| < 1 \right\}$$

in Theorem 3.4. Then

$$\frac{K_0}{8} = \frac{e}{8} < 1.$$

Suppose that y solves

$$(3.9)_\lambda \qquad \begin{cases} y''(t) = -\lambda e^{y(t)}, & t \in [0,1], \\ y(0) = y(1) = 0 \end{cases}$$

for some $\lambda \in (0, 1)$. Then

$$y(t) = -\lambda \int_0^1 G(t,s) e^{y(s)}\, ds$$

and therefore

$$|y(t)| \leq \frac{1}{8} e^{|y|_0} \text{ for } t \in [0,1].$$

Consequently $|y|_0 \leq \frac{1}{8} e^{|y|_0}$ and this implies that $|y|_0 \neq 1$ and therefore $y \notin \partial U$. Now Theorem 3.4 implies that (3.8) has a unique solution with norm *at most* 1.

Remark 3.2 It is interesting to note that (3.8) actually has two solutions. We will discuss multiplicity of solutions in Chapter 7.

Notes The results in Chapter 3 may be found in Frigon [68], Granas [86], Lee and O'Regan [117] and O'Regan [140].

Exercises

3.1 Let X be a Banach space, U an open subset of X with $0 \in U$ and $F : \overline{U} \to X$ a contractive map with $F(\overline{U})$ bounded. Assume that for every $x \in \partial U$, *any one* of the following conditions is satisfied:

(i) $\|F(x)\| \leq \|x\|$,
(ii) $\|F(x)\| \leq \|x - F(x)\|$,
(i) $\|F(x)\| \leq \{\|x\|^2 + \|x - F(x)\|^2\}^{\frac{1}{2}}$,
(iiii) $\|F(x)\| \leq \max\{\|x\|, \|x - F(x)\|\}$,
(iv) $-x \in \overline{U}$ and $F(x) = -F(-x)$.

Show that F has a unique fixed point in U.

3.2 Let X be a Banach space and $F : X \to X$ be α-contractive (i.e. a contraction with Lipschitzian constant α, $0 \leq \alpha < 1$) on every ball $B(0, r)$ in X, and let

$$\Omega_F = \{x \in X : x = \lambda F(x) \text{ for some } \lambda \in (0, 1)\}.$$

Show that either Ω_F is unbounded or F has a fixed point.

3.3 If in Definition 3.1 we replace (d) with

$$(\text{d}^\star) \qquad \begin{cases} \text{there exists a continuous } \phi : [0,1] \to \mathbf{R} \text{ such that} \\ \text{for every } x \in \overline{U} \text{ and } t, s \in [0,1], \text{ we have} \\ d(H(x,t), H(x,s)) \leq |\phi(t) - \phi(s)|, \end{cases}$$

show the result in Theorem 3.1 is again true.

3.4 Let (X, d) be a complete metric space and U an open subset of X. A function $F : \overline{U} \to X$ is said to be *weakly contractive* if there exists $\psi : X \times X \to (0, \infty)$ compactly positive (that is, $\inf\{\psi(x, y) : a \leq d(x, y) \leq b\} = \theta(a, b) > 0$ for every $0 < a \leq b$) such that

$$d(F(x), F(y)) \leq d(x, y) - \psi(x, y).$$

If ψ is a compactly positive function, we define for $0 < a \leq b$,

$$\gamma(a, b) = \min\{a, \theta(a, b)\}.$$

Suppose $x_0 \in X$, $r > 0$ and $F : \overline{B(x_0, r)} \to X$ is weakly contractive with

$$d(x_0, F(x_0)) < \gamma\left(\frac{r}{2}, r\right).$$

Show that F has a fixed point.

3.5 Let (X, d) be a complete metric space, U an open subset of X and $F : \overline{U} \to X$, $G : \overline{U} \to X$ two weakly contractive maps. We say that F and G are homotopic if there exists $H : \overline{U} \times [0, 1] \to X$ such that

(a) $H(\cdot, 0) = G$ and $H(\cdot, 1) = F$,
(b) $x \neq H(x,t)$ for every $x \in \partial U$ and $t \in [0,1]$,
(c) there exists a compactly positive function $\psi : X \times X \to (0, \infty)$ such that $d(H(x,t), H(y,t)) \leq d(x,y) - \psi(x,y)$ for every $x, y \in \overline{U}$ and $t \in [0,1]$,
(d) there exists a continuous $\phi : [0,1] \to \mathbf{R}$ such that for every $x \in \overline{U}$ and $t, s \in [0,1]$, we have that $d(H(x,t), H(x,s)) \leq |\phi(t) - \phi(s)|$.

Now suppose that $F : \overline{U} \to X$ and $G : \overline{U} \to X$ are two weakly contractive maps with *one* of the following conditions satisfied:

(i) $F(\overline{U})$ is bounded;
(ii) there exists a homotopy H between F and G such that the positively compact function ψ associated with H satisfies $\inf\{\theta(a,b) : b \geq a\} > 0$ for all $a > 0$.

If G has a fixed point in U show that F has a fixed point in U.

4
The Theorems of Brouwer, Schauder and Mönch

In this chapter we present fixed point theory for continuous, single valued maps in finite and infinite dimensional Banach spaces. In particular, we present the theorems of Brouwer, Schauder and Mönch.

Definition 4.1 Two topological spaces X and Y are called *homeomorphic* if there exists a bijective function $f : X \to Y$ such that f and f^{-1} are continuous. The map f is called a *homeomorphism.*

Definition 4.2 A topological space X has the *fixed point property* if every continuous $f : X \to X$ has a fixed point.

Theorem 4.1 *If X has the fixed point property and X is homeomorphic to Y, then Y has the fixed point property.*

Proof Let $h : X \to Y$ be a homeomorphism and suppose that $g : Y \to Y$ is continuous. We must show that g has a fixed point in Y. Notice that

$$h^{-1} \circ g \circ h : X \to X$$

is continuous. Since X has the fixed point property, there exists $x_0 \in X$ with

$$h^{-1} \circ g \circ h(x_0) = x_0.$$

Hence $g(y_0) = y_0$ where $y_0 = h(x_0)$. □

Definition 4.3 A subset A of a topological space X is a *retract* of X if there is a continuous map $r : X \to A$ with $r(a) = a$, for all $a \in A$. The map r is called a *retraction.*

Theorem 4.2 *If X has the fixed point property and A is a retract of X, then A has the fixed point property.*

Proof Let $f : A \to A$ be continuous and $r : X \to A$ a retraction. We must show that f has a fixed point in A. Notice first that

$$f \circ r : X \to A \subseteq X.$$

Since X has the fixed point property, there exists $x_0 \in X$ with

$$f \circ r(x_0) = x_0.$$

However, $f(r(x_0)) \in A$ and therefore $x_0 \in A$. But since $x_0 \in A$ and $r : X \to A$ is a retraction, we have that $r(x_0) = x_0$. Consequently, $f(x_0) = x_0$, $x_0 \in A$. □

In $\mathbf{R}$, notice that $f : [-1, 1] \to [-1, 1]$ has a fixed point since the function $g : x \mapsto x - f(x)$ satisfies $g(-1) \leq 0 \leq g(1)$ and therefore g must assume the value zero. As a result, $[-1, 1]$ has the fixed point property. Of course Theorem 4.1 immediately guarantees that all compact intervals have the fixed point property. However, the situation in $\mathbf{R}^n$, $n > 1$, is not as straightforward. A substantial part of this chapter will be devoted to proving the following fixed point result due to Brouwer.

Theorem 4.3 *The closed unit ball B^n, in $\mathbf{R}^n$, has the fixed point property.*

For the remainder of this chapter we shall assume that $\mathbf{R}^n$ is endowed with its standard inner product

$$\langle x, y \rangle = \sum_{i=1}^{n} x_i y_i,$$

and norm

$$\|x\| = \langle x, x \rangle^{1/2}.$$

Also, B^n and S^{n-1} will denote respectively, the closed unit ball and unit sphere in $\mathbf{R}^n$:

$$B^n := \{x \in \mathbf{R}^n : \|x\| \leq 1\} \quad \text{and} \quad S^{n-1} := \{x \in \mathbf{R}^n : \|x\| = 1\}.$$

Definition 4.4 Let $A \subseteq \mathbf{R}^n$. A continuous map $f : A \to \mathbf{R}^n$ is said to be *of class C^1*, if it has a continuous extension to an open neighbourhood of A on which it is continuously differentiable.

In the process of proving Theorem 4.3 we need the following two well known results which we now just state. The proofs are not difficult and can be readily found in the literature (see for example [77]).

Theorem 4.4 *Let A be a compact subset of $\mathbf{R}^n$ and $f : A \to \mathbf{R}^n$ of class C^1 on A. Then there exists a constant $L \geq 0$ (Lipschitz constant) with*

$$\|f(x) - f(y)\| \leq L\|x - y\| \text{ for all } x,\ y \in A.$$

Theorem 4.5 *Let A be a closed, bounded domain (that is, the closure of a connected, open set) in $\mathbf{R}^n$ and $F : A \to \mathbf{R}^n$ of class C^1 on A. Then there exists an interval $(-\epsilon, \epsilon)$, for some $\epsilon > 0$, on which the function*

$$\phi : t \mapsto \mathrm{Vol}(f_t(A))$$

is a polynomial of degree at most n; here $f_t : A \to \mathbf{R}^n$ is given by

$$f_t(x) := x + tF(x),$$

and Vol *refers to the n dimensional Lebesgue measure volume in $\mathbf{R}^n$.*

Definition 4.5 Let $A \subseteq \mathbf{R}^n$. A mapping (vector field) $f : A \to \mathbf{R}^n$ is said to be *nonvanishing* if

$$f(x) \neq 0, \text{ for all } x \in A,$$

and *normed* if

$$\|f(x)\| = 1, \text{ for all } x \in A.$$

The field $f : S^{n-1} \to \mathbf{R}^n$ is said to be *tangent* to S^{n-1} if

$$\langle x, f(x) \rangle = 0 \text{ for all } x \in S^{n-1}.$$

Theorem 4.6 *Suppose that $F : S^{n-1} \to \mathbf{R}^n$ is a normed vector field of class C^1 which is tangent to S^{n-1}. Then for $t > 0$ sufficiently small,*

$$f_t(S^{n-1}) = (1 + t^2)^{1/2} S^{n-1};$$

here $f_t : x \mapsto x + tF(x)$.

Proof Let

$$F^\star(x) := \|x\| F\left(\frac{x}{\|x\|}\right) \text{ for } x \in \mathbf{R}^n \backslash \{0\},$$

and

$$A := \{x \in \mathbf{R}^n : 1/2 \leq \|x\| \leq 3/2\}.$$

Assume that $|t| < \min\{1/3, 1/L\}$, where L is the Lipschitz constant of $F^\star$ on A (see Theorem 4.4). Fix $z \in S^{n-1}$ and define $G : A \to \mathbf{R}^n$ by

$$G(x) := z - tF^\star(x).$$

However, since $\|F(y)\| = 1$ for $y \in S^{n-1}$ and $|t| < 1/3$, we have in fact that $G : A \to A$. Moreover, it is easy to verify (using the Lipschitz constant L of $F^\star$ on A and $|t|L < 1$) that $G : A \to A$ is a contraction. Thus Theorem 1.1 guarantees that G has a fixed point, say $x \in A$, and

$$x + tF^\star(x) = z.$$

Now $1 = \langle x + tF^\star(x), x + tF^\star(x)\rangle$, together with $\|F(u)\| = 1$ for $u \in S^{n-1}$ and $\langle v, F(v)\rangle = 0$ for $v \in S^{n-1}$, immediately gives

$$\|x\| = (1 + t^2)^{-1/2}.$$

As a result

$$y = (1 + t^2)^{1/2} x \in S^{n-1}$$

and therefore

$$y + tF(y) = (1 + t^2)^{1/2} z.$$

We have thus shown that for $|t| < \min\{1/3, 1/L\}$ and any $z \in S^{n-1}$, there exists $y \in S^{n-1}$ with

$$f_t(y) = (1 + t^2)^{1/2} z.$$

Consequently,

$$f_t(S^{n-1}) \subseteq (1 + t^2)^{1/2} S^{n-1}.$$

To show the reverse inclusion, fix $w \in (1 + t^2)^{1/2} S^{n-1}$ and $|t| < \min\{1/3, 1/L\}$. Then there exists $z \in S^{n-1}$ with $w = (1+t^2)^{1/2} z$. From above, for this $z \in S^{n-1}$, there exists $u \in S^{n-1}$ with $f_t(u) = (1+t^2)^{1/2} z$. Thus $f_t(u) = w$ and therefore

$$(1 + t^2)^{1/2} S^{n-1} \subseteq f_t(S^{n-1}). \qquad \square$$

Before we prove Theorem 4.3 we will first show that there are no nonvanishing, continuous vector fields tangent to S^{2n}. Along the way we will need the following result.

Theorem 4.7 *Let $k \in \{1, 2, \ldots\}$. Then there are no normed vector fields of class C^1 tangent to S^{2k}.*

Proof Suppose such a vector field F does exist. Let $0 < a < 1 < b$ and extend F, as in the proof of Theorem 4.6, to $F^\star$ defined on the domain

$$A := \{x \in \mathbf{R}^n : a \leq \|x\| \leq b\}.$$

It is easy to see, since F is tangent to S^{2k}, that $F^\star$ is tangent to *any* sphere concentric with S^{2k} which is contained in A. Let $f_t(x) := x + tF^\star(x)$ and we immediately have from Theorem 4.6 that

$$f_t(A) = (1 + t^2)^{1/2} A,$$

for $t > 0$ sufficiently small. Consequently

$$\mathrm{Vol}(f_t(A)) = (1 + t^2)^{\frac{2k+1}{2}} \mathrm{Vol}(A).$$

However, $(1 + t^2)^{\frac{2k+1}{2}}$ does not coincide with any polynomial in a neighbourhood of zero and this contradicts the conclusion of Theorem 4.5. $\square$

Theorem 4.8 *Let $k \in \{1, 2, \ldots\}$ be fixed. There are no nonvanishing, continuous vector fields tangent to S^{2k}.*

Proof Suppose such a field F exists. Note that

$$m = \min\{\|F(x)\| : x \in S^{2k}\} > 0.$$

By the Weierstrass approximation theorem (applied to each component of F) there exists a vector field $P : S^{2k} \to \mathbf{R}^{2k+1}$ for which

$$\|P(x) - F(x)\| < \frac{m}{2} \text{ for all } x \in S^{2k},$$

where each component of P is a polynomial. Now the field P is of class C^∞ and P is nonvanishing since

$$\|P(x)\| \geq \|F(x)\| - \|P(x) - F(x)\| > \frac{m}{2}, \text{ for } x \in S^{2k}.$$

Define the vector field Q by

$$Q(x) := P(x) - \langle P(x), x \rangle\, x, \text{ for } x \in S^{2k}.$$

Note that Q is of class C^∞ and is easily seen to be tangent to S^{2k}. In addition, for $x \in S^{2k}$ we have

$$\|Q(x)\| \;\geq\; \|P(x)\| - \|Q(x) - P(x)\| > \frac{m}{2} - |\langle P(x), x \rangle|$$

$$= \frac{m}{2} - |\langle P(x) - F(x), x\rangle| \geq \frac{m}{2} - \|P(x) - F(x)\| > 0,$$

since we assumed that F is a nonvanishing vector field, tangent to S^{2k}. Replacing $Q(x)$ with $Q(x)/\|Q(x)\|$, we contradict Theorem 4.7. □

$\mathbf{R}^n$ can be viewed as a subspace of $\mathbf{R}^{n+1}$ by identifying each point $x = (x_1, \ldots, x_n) \in \mathbf{R}^n$ with the point $(x_1, \ldots, x_n, 0) \in \mathbf{R}^{n+1}$. Any point of $\mathbf{R}^{n+1}$ may be represented as (x, x_{n+1}), with $x \in \mathbf{R}^n$ and $x_{n+1} \in \mathbf{R}$. The unit sphere $S^n \subseteq \mathbf{R}^{n+1}$ may be divided into the upper hemisphere

$$S^n_+ := \{(x, x_{n+1}) \in S^n : x_{n+1} \geq 0\},$$

and the lower hemisphere

$$S^n_- := \{(x, x_{n+1}) \in S^n : x_{n+1} \leq 0\}.$$

The unit sphere

$$S^{n-1} := S^n_+ \cap S^n_-$$

is the equator. Let

$e_{n+1} := (0, \ldots, 0, 1)$ (north pole) and $-e_{n+1} := (0, \ldots, 0, -1)$ (south pole).

The well known stereographic projection (from e_{n+1} to S^n) is the mapping $S_+ : \mathbf{R}^n \to S^n$ defined by

$$S_+(x) := \left(\frac{2x}{1 + \|x\|^2}, \frac{\|x\|^2 - 1}{1 + \|x\|^2}\right), \quad \text{for } x \in \mathbf{R}^n.$$

This map is of class C^∞ and it transforms B^n onto S^n_-. In addition, $S_+(x) = x$ for $x \in S^{n-1}$. We define the stereographic projection S_- (from $-e_{n+1}$ to S^n) by

$$S_-(x) := \left(\frac{2x}{1 + \|x\|^2}, \frac{1 - \|x\|^2}{1 + \|x\|^2}\right), \quad \text{for } x \in \mathbf{R}^n.$$

Proof of Theorem 4.3 First we assume that $n = 2k$. Suppose there exists a continuous map $f : B^{2k} \to B^{2k}$ which has no fixed points. Define the vector field G by

$$G(x) := x - f(x).$$

It is immediate that G is nonvanishing on $\mathbf{B}^{2k}$, and it is easy to see that at any point $x \in S^{2k-1}$ the field is directed outwards, that is,

$$\langle G(x), x\rangle = 1 - \langle x, f(x)\rangle > 0 \text{ for } x \in S^{2k-1}.$$

Now let

$$F(x) := x - \left(\frac{1 - \|x\|^2}{1 - \langle x, f(x) \rangle} \right) f(x).$$

If $F(x) = 0$ for $x \in B^{2k}$, then 0, x and $f(x)$ are collinear, and therefore

$$\langle x, f(x) \rangle \, x = \|x\|^2 f(x).$$

This in turn immediately implies that $x = f(x)$ since $F(x) = 0$ – a contradiction. Thus F is nonvanishing on B^{2k} and it is also clear that $F(x) = x$ if $x \in S^{2k-1}$.

For any $x \in B^{2k}$, consider the set $\{x + tF(x) : t \in [0,1]\}$. The image of this set under S_+ is a differentiable arc with initial point lying on S^{2k}_-. Define the vector field T_- on S^{2k}_- by

$$\begin{aligned} T_-(y) &:= \left\{ \frac{d}{dt} S_+(x + tF(x)) \right\} \bigg|_{t=0} \\ &= \frac{2}{(1 + \|x\|^2)^2} \left((1 + \|x\|^2) F(x) - 2 \langle x, F(x) \rangle \, x, \, 2 \langle x, F(x) \rangle \right), \end{aligned}$$

for $y = S_+(x) \in S^{2k}_-$. It is easy to see (since $F(x) \neq 0$ for $x \in B^{2k}$) that T_- is nonvanishing, continuous and tangent to S^{2k}_-. In addition, $F(x) = x$ for $x \in S^{2k-1}$ implies $T_-(y) = e_{n+1}$ for $y \in S^{2k-1}$.

Define the vector field T_+ on S^{2k}_+ by

$$\begin{aligned} T_+(y) &:= \left\{ \frac{d}{dt} S_-(x - tF(x)) \right\} \bigg|_{t=0} \\ &= \frac{2}{(1 + \|x\|^2)^2} \left(2 \langle x, F(x) \rangle \, x - (1 + \|x\|^2) F(x), \, 2 \langle x, F(x) \rangle \right), \end{aligned}$$

for $y = S_-(x) \in S^{2k}_+$. It is easy to see that $T_+(y) = T_-(y)$ for $y \in S^{2k-1}$.

Now for $y \in S^{2k}$, set

$$T(y) := \begin{cases} T_-(y) \text{ for } y \in S^{2k}_-, \\ T_+(y) \text{ for } y \in S^{2k}_+. \end{cases}$$

Then T is a continuous, nonvanishing vector field tangent to S^{2k}. This contradicts Theorem 4.8. Consequently, f has a fixed point and therefore our proof is complete if $n = 2k$.

To prove the result for $B^n = B^{2k-1}$, it suffices to observe that if the continuous mapping $f : B^{2k-1} \to B^{2k-1}$ is fixed point free, then so also is $g : B^{2k} \to B^{2k}$ defined by

$$g(x, x_{2k}) := (f(x), 0).$$

This contradicts the first part of the proof. □

In fact it is now easy to prove a generalisation of Theorem 4.3 if we note the following result.

Theorem 4.9 *Every nonempty, closed, convex subset C of $\mathbf{R}^n$ is a retract of $\mathbf{R}^n$.*

Proof For any $x \in \mathbf{R}^n$, we know that there exists a unique $y = P_C(x) \in C$ with

$$\|x - y\| = \inf\{\|x - u\| : u \in C\},$$

that is, P_C is the map sending each $x \in H$ to the nearest point in C. From Exercise 2.5, we know that P_C is nonexpansive, therefore in particular, a retraction from $\mathbf{R}^n$ onto C. □

Theorem 4.10 *Every nonempty, bounded, closed, convex subset C of $\mathbf{R}^n$ has the fixed point property.*

Proof Notice that C is a subset of some ball $B^\star$ in $\mathbf{R}^n$. Since B^n and $B^\star$ are homeomorphic, Theorem 4.1 and Theorem 4.3 guarantee that $B^\star$ has the fixed point property. In addition, Theorem 4.9 implies that C is a retract of $B^\star$ and therefore Theorem 4.2 ensures that C has the fixed point property. □

Remark 4.1 Since any finite dimensional normed linear space X is isomorphic to $\mathbf{R}^n$ with $n = \dim X$ we have: every nonempty, bounded, closed, convex subset of a finite dimensional normed linear space has the fixed point property.

We would like to extend Theorem 4.10 to an infinite dimensional space setting. To do so, additional assumptions (see the following example) have to be placed on f.

Example 4.1 Let

$$l_2 := \left\{ x = (x_1, x_2, \ldots) : \|x\|^2 = \sum_{i=1}^{\infty} |x_i|^2 < \infty \right\}$$

and

$$B := \{x \in l_2 : \|x\| \leq 1\}.$$

Define $f : B \to \partial B \subseteq B$ by

$$f(x) := \left(\sqrt{1 - \|x\|^2}, x_1, x_2, \ldots\right).$$

It is easy to see that f is continuous but does not have a fixed point.

Definition 4.6 Let X and Y be normed linear spaces. A map $F : X \to Y$ is called *compact* if $F(X)$ is contained in a compact subset of Y. A compact map $F : X \to Y$ is called *finite dimensional,* if $F(X)$ is contained in a finite dimensional linear subspace of Y.

We next extend Brouwer's fixed point theorem to compact maps in normed linear spaces. This generalisation is due to Schauder. The main idea is to approximate compact maps by maps with finite dimensional ranges.

Let $A = \{a_1, \ldots, a_n\}$ be a finite subset of a normed linear space $E = (E, \|\cdot\|)$ and for fixed $\epsilon > 0$ let

$$A_\epsilon := \bigcup_{i=1}^{n} B(a_i, \epsilon) \text{ where } B(a_i, \epsilon) := \{x \in E : \|x - a_i\| < \epsilon\}.$$

For each $i = 1, \ldots, n$, let $\mu_i : A_\epsilon \to \mathbf{R}$ be the map given by

$$\mu_i(x) := \max\{0, \epsilon - \|x - a_i\|\}.$$

Let $\mathrm{co}(A)$ denote the smallest convex set containing A. The *Schauder projection* is the map $P_\epsilon : A_\epsilon \to \mathrm{co}(A)$ given by

$$P_\epsilon(x) := \frac{\sum_{i=1}^{n} \mu_i(x) a_i}{\sum_{i=1}^{n} \mu_i(x)} \quad \text{for } x \in A_\epsilon.$$

Notice $P_\epsilon(x)$ is well defined since if $x \in A_\epsilon$, then $x \in B(a_i, \epsilon)$ for some $i \in \{1, 2, \ldots\}$ and therefore $\sum_{i=1}^{n} \mu_i(x) \neq 0$. Also $P_\epsilon(x) \subseteq \mathrm{co}(A)$ since each $P_\epsilon(x)$ is a convex combination of the points $a_1, \ldots, a_n$.

Theorem 4.11 *Let C be a convex subset of a normed linear space, and $A = \{a_1, \ldots, a_n\} \subseteq C$. If P_ϵ denotes the Schauder projection, then*

(i) *P_ϵ is a compact, continuous map from A_ϵ into $\mathrm{co}(A) \subseteq C$, and*
(ii) *$\|x - P_\epsilon(x)\| < \epsilon$ for all $x \in A_\epsilon$.*

Proof

(i) The continuity of P_ϵ is immediate. To show compactness, let $\{P_\epsilon(x_m)\}_{m=1}^{\infty}$ be any sequence in $P_\epsilon(A_\epsilon)$. Let $\mu(x) := \sum_{i=1}^{n} \mu_i(x)$

and therefore

$$P_\epsilon(x_m) := \sum_{i=1}^{n} \frac{\mu_i(x_m)}{\mu(x_m)} a_i.$$

Notice that for each $m \in \{1, 2, \ldots\}$,

$$\left(\frac{\mu_1(x_m)}{\mu(x_m)}, \ldots, \frac{\mu_n(x_m)}{\mu(x_m)} \right) \in [0,1]^n,$$

therefore the compactness of the n-cube implies the compactness of the map P_ϵ.

(ii) Notice that for $x \in A_\epsilon$,

$$\begin{aligned} \|x - P_\epsilon(x)\| &= \frac{1}{\mu(x)} \left\| \mu(x)x - \sum_{i=1}^{n} \mu_i(x) a_i \right\| \\ &\leq \frac{1}{\mu(x)} \sum_{i=1}^{n} \mu_i(x) \|x - a_i\| < \frac{1}{\mu(x)} \sum_{i=1}^{n} \mu_i(x) \epsilon = \epsilon, \end{aligned}$$

since $\mu_i(x) = 0$ unless $\|x - a_i\| < \epsilon$. □

Our next result is known as Schauder's approximation theorem.

Theorem 4.12 *Let C be a convex subset of a normed linear space E and $F : E \to C$ a compact, continuous map. Then for each $\epsilon > 0$, there are a finite set $A = \{a_1, \ldots, a_n\}$ in $F(E)$ and a finite dimensional continuous map $F_\epsilon : E \to C$ with the following properties:*

(i) $\|F_\epsilon(x) - F(x)\| < \epsilon$ *for all* $x \in E$,

(ii) $F_\epsilon(x) \subseteq \mathrm{co}(A) \subseteq C$.

Proof $F(E)$ is contained in a compact subset K of C, therefore since K is totally bounded, there exists a set $\{a_1, \ldots, a_n\} \subseteq F(E)$ with $F(E) \subseteq A_\epsilon$. Let $P_\epsilon : A_\epsilon \to \mathrm{co}(A)$ be the Schauder projection and define the map $F_\epsilon : E \to C$ by

$$F_\epsilon(x) := P_\epsilon(F(x)) \text{ for } x \in E.$$

Theorem 4.11 now guarantees the result. □

Before we prove Schauder's fixed point theorem we first introduce the notion of an ϵ-fixed point (see Exercise 2.3). Let D be a subset of a normed linear space E and $F : D \to E$ a map. Given $\epsilon > 0$, a point $d \in D$ with $\|d - F(d)\| < \epsilon$ is called an *ϵ-fixed point* for F.

Theorem 4.13 *Let D be a closed subset of a normed linear space E and $F : D \to E$ a compact, continuous map. Then F has a fixed point if and only if F has an ϵ-fixed point.*

Proof Assume that F has an ϵ-fixed point for each $\epsilon > 0$. Now for each $n \in \{1, 2, \ldots\}$, let d_n be a $(1/n)$-fixed point for F, that is,

$$\|d_n - F(d_n)\| < \frac{1}{n}. \tag{4.1}$$

Since F is compact, $F(D)$ is contained in a compact subset K of E and therefore there exist a subsequence S of integers and an $x \in K$ such that

$$F(d_n) \to x \in K \text{ as } n \to \infty \text{ in } S.$$

Now (4.1) implies that $d_n \to x$ as $n \to \infty$ in S and since D is closed we have that $x \in D$. Also, the continuity F implies that $F(d_n) \to F(x)$ as $n \to \infty$ in S and this together with (4.1) yields $\|x - F(x)\| = 0$, that is $x = F(x)$. □

We now state and prove Schauder's fixed point theorem.

Theorem 4.14 *Let C be a closed, convex subset of a normed linear space E. Then every compact, continuous map $F : C \to C$ has at least one fixed point.*

Proof By Theorem 4.13, with $D = C$, it suffices to show that F has an ϵ-fixed point for every $\epsilon > 0$. Fix $\epsilon > 0$. Theorem 4.12 guarantees the existence of a finite dimensional, continuous map $F_\epsilon : C \to C$ with

$$\|F_\epsilon(x) - F(x)\| < \epsilon \text{ for } x \in C \tag{4.2}$$

and

$$F_\epsilon(C) \subseteq \text{co}(A) \subseteq C \text{ for some finite set } A \subseteq C.$$

Since $\text{co}(A)$ is closed and bounded and $F_\epsilon(\text{co}(A)) \subseteq \text{co}(A)$, we may apply Theorem 4.10 (Brouwer's fixed point theorem) to deduce that there exists $x_\epsilon \in \text{co}(A)$ with $x_\epsilon = F_\epsilon(x_\epsilon)$. Also, (4.2) yields

$$\|x_\epsilon - F_\epsilon(x_\epsilon)\| = \|F_\epsilon(x_\epsilon) - F(x_\epsilon)\| < \epsilon. \qquad \square$$

In 1980, Mönch generalised Theorem 4.14 and for completeness we present his result here.

Theorem 4.15 *Let Ω be a convex, open set in a Banach space E with $x_0 \in \Omega$. Suppose there is a continuous map $F : \overline{\Omega} \to \overline{\Omega}$ with the following property:*

$$(4.3) \qquad \begin{cases} C \subseteq \overline{\Omega} \text{ countable and } C \subseteq \overline{\mathrm{co}}(\{x_0\} \cup F(C)) \\ \text{imply that } C \text{ is relatively compact.} \end{cases}$$

Then F has a fixed point in $\overline{\Omega}$.

Proof Let

$$D_0 := \{x_0\} \text{ and } D_n := \mathrm{co}(\{x_0\} \cup F(D_{n-1})) \text{ for } n = 1, 2, \ldots.$$

Mazur's theorem (Exercise 4.12), together with the fact that the continuous image of a compact set is compact, implies that D_n is relatively compact. Notice also that

$$D_0 \subseteq D_1 \subseteq \cdots \subseteq D_{n-1} \subseteq D_n \subseteq \cdots \subseteq \overline{\Omega}.$$

In addition, each D_n is separable, therefore there is a sequence of countable $\{C_n\}$, with $\overline{C_n} = \overline{D_n}$ for $n = 0, 1, 2, \ldots$. Let

$$D := \bigcup_{n=0}^{\infty} D_n \text{ and } C := \bigcup_{n=0}^{\infty} C_n.$$

Notice it is easy to check since (D_n) is increasing, that

$$(4.4) \qquad D = \bigcup_{n=0}^{\infty} D_n = \bigcup_{n=1}^{\infty} \mathrm{co}(\{x_0\} \cup F(D_{n-1})) = \mathrm{co}(\{x_0\} \cup F(D))$$

and $\left(\text{note } \bigcup_{n=0}^{\infty} D_n \subseteq \bigcup_{n=0}^{\infty} \overline{D_n} \subseteq \overline{\bigcup_{n=0}^{\infty} D_n}\right)$

$$(4.5) \qquad \overline{\bigcup_{n=0}^{\infty} \overline{D_n}} = \overline{\bigcup_{n=0}^{\infty} D_n} = \overline{D} \text{ and } \overline{\bigcup_{n=0}^{\infty} \overline{D_n}} = \overline{\bigcup_{n=0}^{\infty} \overline{C_n}} = \overline{\bigcup_{n=0}^{\infty} C_n} = \overline{C}.$$

It follows from (4.4), (4.5) and the continuity of F (that is, $F(\overline{D}) \subseteq \overline{F(D)}$), that

$$(4.6) \qquad \begin{aligned} C \subseteq \overline{C} = \overline{D} &= \overline{\mathrm{co}}(\{x_0\} \cup F(D)) = \overline{\mathrm{co}}(\{x_0\} \cup F(\overline{D})) \\ &= \overline{\mathrm{co}}(\{x_0\} \cup F(\overline{C})) = \overline{\mathrm{co}}(\{x_0\} \cup F(C)). \end{aligned}$$

Remark 4.2 Note that the continuity of F implies that

$$F(D) \cup \{x_0\} \subseteq F(\overline{D}) \cup \{x_0\} \subseteq \overline{F(D) \cup \{x_0\}} \subseteq \overline{\mathrm{co}}(F(D) \cup \{x_0\})$$

and therefore

$$\overline{co}(F(D) \cup \{x_0\}) = \overline{co}(F(\overline{D}) \cup \{x_0\}).$$

Since C is countable, (4.3) implies that $\overline{C}$ is compact. Hence $\overline{D}$ is compact. Also notice that $\overline{D} = \overline{co}(\{x_0\} \cup F(\overline{D}))$ from (4.6) and therefore $F(\overline{D}) \subseteq \overline{D}$. We may apply Theorem 4.14 (Schauder's fixed point theorem) to deduce that F has a fixed point in $\overline{D}$. □

We are now in a position to prove Mönch's fixed point theorem.

Theorem 4.16 *Let C be a closed, convex subset of a Banach space E with $x_0 \in C$. Suppose there is a continuous map $F : C \to C$ with the following property:*

$$(4.7) \qquad \begin{cases} D \subseteq C \text{ countable and } D \subseteq \overline{co}(\{x_0\} \cup F(D)) \\ \text{imply that } D \text{ is relatively compact.} \end{cases}$$

Then F has a fixed point in C.

Proof Essentially the same reasoning as in Theorem 4.15 establishes this. However here we provide a second proof. By Dugundji's extension theorem (see Exercise 4.8) there is a continuous map $F^\star : E \to C$ with $F^\star(x) = F(x)$ for $x \in C$. Let $D \subseteq E$ be countable and

$$(4.8) \qquad D \subseteq \overline{co}(\{x_0\} \cup F^\star(D)).$$

Since $F^\star(D) \subseteq C$, (4.8) implies that $D \subseteq C$ and as a result $F^\star(D) = F(D)$. We may apply (4.7) to deduce that D is relatively compact. Next apply Theorem 4.15 to $F^\star$ with $\Omega = E$ to show that there exists $x \in E$ with $F^\star(x) = x$. Since $F^\star(x) \in C$ we must have $x \in C$ and therefore $x = F^\star(x) = F(x)$. □

Remark 4.3 Theorem 4.16 contains the well known fixed point theorem of Sadovskii (see Exercise 4.5).

To illustrate how Theorem 4.14 can be applied in practice, we discuss the discrete boundary value problem

$$(4.9) \qquad \begin{cases} \Delta^2 y(i-1) + f(i, y(i)) = 0, \ \ i \in N, \\ y(0) = 0, \ \ y(T+1) = 0. \end{cases}$$

Here $T \in \{1, 2, \ldots\}$, $N = \{1, 2, \ldots, T\}$, $N^+ = \{0, 1, \ldots, T+1\}$, $\Delta y(j) = y(j+1) - y(j)$, and $y : N^+ \to \mathbf{R}^n$. We will assume that

$$f : N \times \mathbf{R}^n \to \mathbf{R}^n \text{ is continuous} \tag{4.10}$$

holds.

Remark 4.4 Recall that a map $f : N^+ \times \mathbf{R}^n \to \mathbf{R}^n$ is continuous as a map of the topological space $N \times \mathbf{R}^n$ into the topological space $\mathbf{R}^n$ (the topology on N will be the discrete topology).

Let $C(N^+, \mathbf{R}^n)$ denote the class of maps w, continuous on N^+ (discrete topology), with norm $\|w\|_0 = \max_{k \in N^+} \|w(k)\|$, that is, $C(N^+, \mathbf{R}^n) := \{w : w : N^+ \to \mathbf{R}^n\}$ which is a Banach space.

Remark 4.5 Since N^+ is a discrete space, then any mapping of N^+ to a topological space (in this case $\mathbf{R}^n$) is continuous.

By a solution of (4.9) we mean a $w \in C(N^+, \mathbf{R}^n)$ such that w satisfies (4.9) for $i \in N$ and w satisfies the boundary (Dirichlet) conditions. We will establish two existence results for (4.9). Both results will depend on the following result (here $\langle \cdot, \cdot \rangle$ is as defined in the beginning of this chapter).

Theorem 4.17 *Let $x \in C(N^+, \mathbf{R}^n)$ and $i \in \{0, 1, \ldots, T-1\}$. Then*

$$\|x(i+1)\| \Delta^2 \|x(i)\| \geq \left\langle x(i+1), \Delta^2 x(i) \right\rangle.$$

Proof Notice that

$$\begin{aligned}
\|x(i+1)\| \Delta^2 \|x(i)\| \quad - \quad & \langle x(i+1), \Delta^2 x(i) \rangle \\
= \quad & (\|x(i+2)\| - 2\|x(i+1)\| + \|x(i)\|)\, \|x(i+1)\| \\
& - \langle x(i+1),\, x(i+2) - 2x(i+1) + x(i) \rangle \\
= \quad & \|x(i+2)\|\, \|x(i+1)\| + \|x(i)\|\, \|x(i+1)\| \\
& - \langle x(i+1),\, x(i+2) + x(i) \rangle \\
\geq \quad & \|x(i+2)\|\, \|x(i+1)\| + \|x(i)\|\, \|x(i+1)\| \\
& - \|x(i+1)\|\, (\|x(i+2)\| + \|x(i)\|) = 0.
\end{aligned}$$

□

Theorem 4.18 *Suppose that (4.10) holds. In addition, assume the following hold:*

$$\text{(4.11)}\quad \begin{cases} \text{there exist } v \in C(N^+, \mathbf{R}^n) \text{ and } M \in C(N^+, (0,\infty)) \text{ with} \\ \langle y - v(i), -f(i,y) - \Delta^2 v(i-1)\rangle \geq M(i)\Delta^2 M(i-1) \\ \text{for all } i \in N \text{ and } y \in \mathbf{R}^n \text{ with } \|y - v(i)\| = M(i) \end{cases}$$

and

$$\text{(4.12)}\qquad \|v(0)\| \leq M(0) \text{ with } \|v(T+1)\| \leq M(T+1).$$

Then (4.9) has a solution $y \in C(N^+, \mathbf{R}^n)$ with $\|y(i) - v(i)\| \leq M(i)$, for $i \in N$.

Remark 4.6 Suppose that $n = 1$ and α, $\beta \in C(N^+, \mathbf{R})$ are respectively lower and upper solutions of (4.9), that is,

$$\begin{cases} \Delta^2\alpha(i-1) + f(i,\alpha(i)) \geq 0 \text{ for } i \in N, \\ \alpha(0) \leq 0, \ \ \alpha(T+1) \leq 0 \end{cases}$$

and

$$\begin{cases} \Delta^2\beta(i-1) + f(i,\beta(i)) \geq 0 \text{ for } i \in N, \\ \beta(0) \geq 0, \ \ \beta(T+1) \geq 0. \end{cases}$$

In addition, suppose that $\alpha(i) < \beta(i)$ for $i \in N$. Then it is easy to check that

$$v = \frac{\alpha + \beta}{2} \quad \text{and} \quad M = \frac{\beta - \alpha}{2}$$

satisfy (4.11) and (4.12).

Proof of Theorem 4.18 Consider the discrete boundary value problem

$$\text{(4.13)}\qquad \begin{cases} \Delta^2 y(i-1) + f(i, P(i, y(i))) = 0, \ \ i \in N, \\ y(0) = 0, \ \ y(T+1) = 0, \end{cases}$$

where

$$P(i,y) := \min\left\{1, \frac{M(i)}{\|y - v(i)\|}\right\} y + \left(1 - \min\left\{1, \frac{M(i)}{\|y - v(i)\|}\right\}\right) v(i).$$

That is,

$$P(i,y) = \begin{cases} y & \text{if } \|y - v(i)\| \leq M(i), \\ M(i)\dfrac{y - v(i)}{\|y - v(i)\|} + v(i) & \text{if } \|y - v(i)\| > M(i) \end{cases}$$

is the radial retraction of $\mathbf{R}^n$ onto $\{y : \|y - v(i)\| \le M(i)\}$. We now use Theorem 4.14 to show that (4.13) has a solution. Solving (4.13) is equivalent to finding a $y \in C(N^+, \mathbf{R}^n)$ which satisfies

$$y(i) = \sum_{j=1}^{T} G(i,j) f(j, P(j, y(j))) \text{ for } i \in N^+, \tag{4.14}$$

where

$$G(i,j) := \begin{cases} \dfrac{j(T+1-i)}{T+1}, & 0 \le j \le i-1, \\ \dfrac{i(T+1-j)}{T+1}, & i \le j \le T+1. \end{cases}$$

Define the operator $F : C(N^+, \mathbf{R}^n) \to C(N^+, \mathbf{R}^n)$ by setting

$$Fy(i) := \sum_{j=1}^{T} G(i,j) f(j, P(j, y(j))).$$

Now (4.14) is equivalent to the fixed point problem $y = F(y)$. It is immediate from the Arzelà–Ascoli theorem in $C(N^+, \mathbf{R}^n)$ that $F : C(N^+, \mathbf{R}^n) \to C(N^+, \mathbf{R}^n)$ is a compact map. Schauder's fixed point theorem (Theorem 4.14) guarantees that F has a fixed point. Consequently, (4.13) has a solution $y \in C(N^+, \mathbf{R}^n)$.

It remains to show that $\|y(i) - v(i)\| \le M(i)$ for $i \in N$. If this is not true then

$$r(i) = \|y(i) - v(i)\| - M(i)$$

attains a positive global maximum at say $m \in N$ and we may assume, without loss of generality, that $r(m) > r(m-1)$. Thus $r(m) > r(m-1)$ and $r(m) \ge r(m+1)$ imply

$$r(m+1) - 2r(m) + r(m-1) < 0.$$

Consequently,

$$\Delta^2 r(m-1) < 0. \tag{4.15}$$

On the other hand, since $r(m) > 0$, we have using Theorem 4.17 and assumption (4.11), that

$$\begin{aligned} \Delta^2 r(m-1) &= \Delta^2 \|y(m-1) - v(m-1)\| - \Delta^2 M(m-1) \\ &\ge \frac{\langle y(m) - v(m), \Delta^2 (y(m-1) - v(m-1)) \rangle}{\|y(m) - v(m)\|} \\ &\quad - \Delta^2 M(m-1) \end{aligned}$$

$$
\begin{aligned}
&= \frac{\langle P(m,y(m)) - v(m), \Delta^2 y(m-1) - \Delta^2 v(m-1)\rangle}{M(m)} \\
&\quad -\Delta^2 M(m-1) \\
&= \frac{\langle P(m,y(m)) - v(m), -f(m,P(m,y(m))) - \Delta^2 v(m-1)\rangle}{M(m)} \\
&\quad -\Delta^2 M(m-1) \\
&\geq \Delta^2 M(m-1) - \Delta^2 M(m-1) = 0.
\end{aligned}
$$

This contradicts (4.15). Thus $\|y(i) - v(i)\| \leq M(i)$ for $i \in N$ and we are finished. □

In fact it is also possible to discuss the case when $M(i)$, in (4.11), may take on the value zero.

Theorem 4.19 *Suppose that* (4.10) *holds. In addition, assume the following hold:*

$$
(4.16) \quad \begin{cases} \text{there exist } v \in C(N^+, \mathbf{R}^n) \text{ and } M \in C(N^+, [0,\infty)) \text{ with} \\ \langle y - v(i), -f(i,y) - \Delta^2 v(i-1)\rangle \geq M(i)\Delta^2 M(i-1) \\ \text{for all } i \in N \text{ and } y \in \mathbf{R}^n \text{ with } \|y - v(i)\| = M(i), \\ M(i) \neq 0, \end{cases}
$$

$$
(4.17) \quad \begin{cases} \text{there exist } v \text{ and } M \text{ as in (4.16) with} \\ \dfrac{\langle y - v(i), -f(i,v(i)) - \Delta^2 v(i-1)\rangle}{\|y - v(i)\|} \geq \Delta^2 M(i-1) \\ \text{for all } i \in N \text{ and } y \in \mathbf{R}^n \text{ with } \|y - v(i)\| > M(i), \\ M(i) = 0, \end{cases}
$$

and

$$
(4.18) \qquad \|v(0)\| \leq M(0) \text{ with } \|v(T+1)\| \leq M(T+1).
$$

Then (4.9) *has a solution* $y \in C(N^+, \mathbf{R}^n)$ *with* $\|y(i) - v(i)\| \leq M(i)$ *for* $i \in N$.

Remark 4.7 If $n = 1$ and α, $\beta \in C(N^+, \mathbf{R}^n)$ are respectively lower and upper solutions of (4.9), with $\alpha(i) \leq \beta(i)$ for $i \in N$, then it is easy to check that

$$
v = \frac{\alpha + \beta}{2} \text{ and } M = \frac{\beta - \alpha}{2}
$$

satisfy (4.16), (4.17) and (4.18).

Proof of Theorem 4.19 As in Theorem 4.18, (4.13) has a solution $y \in C(N^+, \mathbf{R}^n)$. Let $r(i)$ and m be as in Theorem 4.18 and once again (4.15) is true. On the other hand, if $M(m) > 0$, then exactly the same argument as in Theorem 4.18 establishes a contradiction.

Next we see that if $M(m) = 0$, then Theorem 4.17 and assumption (4.17) yield

$$\begin{aligned}
\Delta^2 r(m-1) &\geq \frac{\langle y(m) - v(m), \Delta^2 y(m-1) - \Delta^2 v(m-1)\rangle}{\|y(m) - v(m)\|} \\
&\quad -\Delta^2 M(m-1) \\
&= \frac{\langle y(m) - v(m), -f(m, P(m, y(m))) - \Delta^2 v(m-1)\rangle}{\|y(m) - v(m)\|} \\
&\quad -\Delta^2 M(m-1) \\
&= \frac{\langle y(m) - v(m), -f(m, v(m)) - \Delta^2 v(m-1)\rangle}{\|y(m) - v(m)\|} \\
&\quad -\Delta^2 M(m-1) \geq 0.
\end{aligned}$$

In both cases we contradict (4.15) and therefore the proof is complete. □

Notes We follow the approach of Goebel and Kirk [77] to establish Brouwer's fixed point theorem. Other approaches may be found in Border [23] and Dugundji and Granas [55]. The proof of Schauder's fixed point theorem can be found in any classical fixed point theory book [55, 77, 174, 191]. Mönch's original paper [130] and a recent book of Guo, Lakshmikantham and Liu [88] were used to establish Theorem 4.15 and Theorem 4.16. Theorem 4.18 and Theorem 4.19 were taken from Agarwal and O'Regan [1].

Exercises

4.1 Let E be a normed space, B the closed unit ball in E and ∂B the unit sphere in E. Suppose that F is a continuous, compact map of B into E with $F(\partial B) \subseteq B$. Show that F has a fixed point.

4.2 Let E be a normed space, F a continuous mapping of E into E which is compact on each bounded subset of E. Show that either

(i) the equation $x = \lambda F(x)$ has a solution for $\lambda = 1$, or

(ii) the set of all such solutions x, for $0 < \lambda < 1$, is unbounded holds.

4.3 Let E be a Banach space and Ω_E the class of bounded subsets of E. The Kuratowskii measure of noncompactness $\alpha : \Omega_E \to [0, \infty)$ is defined by (here $x \in \Omega_E$)

$$\alpha(X) := \inf\left\{\epsilon > 0 : X \subseteq \bigcup_{i=1}^{n} X_i \text{ and } \operatorname{diam}(X_i) \leq \epsilon \text{ for } i = 1, \ldots, n\right\};$$

here $\operatorname{diam} X_i = \sup\{|x - y| : x, y \in X_i\}$.

Let A, $B \in \Omega_E$. Show that the following hold:

(i) $\alpha(A) = 0$ if and only if $\overline{A}$ is compact,
(ii) $\alpha(\overline{A}) = \alpha(A)$,
(iii) $A \subseteq B$ implies that $\alpha(A) \leq \alpha(B)$,
(iv) $\alpha(A \cup B) = \max\{\alpha(A), \alpha(B)\}$,
(v) $\alpha(\lambda A) = |\lambda|\alpha(A)$, $\lambda \in \mathbf{R}$,
(vi) $\alpha(A + B) \leq \alpha(A) + \alpha(B)$,
(vii) $\alpha(\operatorname{co}(A)) = \alpha(A)$.

4.4 Let E be a Banach space and $\{A_n\}$ a decreasing sequence of nonempty, closed, bounded subsets of E with $\alpha(A_n) \to 0$. Show that $A = \bigcap_{n=1}^{\infty} A_n$ is nonempty and compact.

4.5 Let C be a closed, convex subset of a Banach space E. Suppose that $F : C \to C$ is a continuous, condensing ($\alpha(F(X)) < \alpha(X)$ for any $X \subseteq C$ with $\alpha(X) \neq 0$) map, with $F(C)$ a bounded set in C. Use Theorem 4.16 to show that F has a fixed point in C.

4.6 Provide a proof of Theorem 4.4.

4.7 Provide a proof of Theorem 4.5.

4.8 Let $T : A \subseteq X \to Y$ be a continuous operator on a nonempty, closed subset A of the metric space (X, d) to the normed space Y. Show that T has a continuous extension $T^\star : X \to \operatorname{co}(T(A))$.

4.9 Let U be a bounded, symmetric with respect to zero (that is, $x \in U$ implies that $-x \in U$), convex open neighbourhood of zero in $\mathbf{R}^n$ with $F : \overline{U} \to \mathbf{R}^n$ continuous and antipodal preserving on ∂U (that is $F(-a) = -F(a)$ for $a \in \partial U$). Show that F has a fixed point.

4.10 Let $N = \{a_1, \ldots, a_n\}$ belong to some convex set C in a normed linear space E and let P_ϵ be the Schauder projection. Show that if $N \subseteq C$ is symmetric with respect to zero ($N = \{a_1, \ldots, a_n, -a_1, \ldots, -a_n\}$), then $A_\epsilon = -A_\epsilon$ and $P_\epsilon(-x) = -P_\epsilon(x)$ for all $x \in A_\epsilon$; here $A_\epsilon = \bigcup_{i=1}^{n} B(a_i, \epsilon)$.

4.11 Let U be a bounded, convex, open, symmetric neighbourhood of zero in a normed linear space E. Show that every compact, continuous map $F : \overline{C} \to E$ that is antipodal preserving on ∂U (that is, $F(-a) = -F(a)$ on ∂U) has at least one fixed point.

4.12 Show that the closed, convex hull of a compact set in a Banach space is compact.

4.13 In Theorem 4.14, show that it is enough to assume C is convex, that is, C need not necessarily be closed.

4.14 In Theorem 4.16 show that (4.7) may be replaced by

$$\begin{cases} D \subseteq C \text{ countable and } \overline{D} = \overline{\text{co}}(\{x_0\} \cup F(D)) \\ \text{imply that } \overline{D} \text{ is compact.} \end{cases}$$

5
Nonlinear Alternatives of Leray–Schauder Type

To apply the Schauder or Mönch fixed point theorem we need to find a closed, convex set that is mapped by the map under investigation back into itself. From an application viewpoint this is extremely difficult to achieve. As a result, we turn our attention in this chapter to establishing a fixed point theory for nonself maps.

Throughout the chapter, E will be a Banach space, C a closed, convex subset of E and U an open subset of C. We discuss maps $F : \overline{U} \to C$, where $\overline{U}$ denotes the closure of U in C. Our first result concerns continuous, compact maps.

Theorem 5.1 *Let E be a Banach space, C a closed, convex subset of E, U an open subset of C and $p \in U$. Suppose that $F : \overline{U} \to C$ is a continuous, compact (that is, $F(\overline{U})$ is a relatively compact subset of C) map. Then either*

(A1) *F has a fixed point in $\overline{U}$, or*

(A2) *there is a $u \in \partial U$ (the boundary of U in C) and $\lambda \in (0,1)$ with $u = \lambda F(u) + (1-\lambda)p$.*

Proof Suppose (A2) does not hold and F has no fixed points on ∂U (otherwise we are finished). Then

$$u \neq \lambda F(u) + (1-\lambda)p \text{ for } u \in \partial U \text{ and } \lambda \in [0,1].$$

Consider

$$A := \{x \in \overline{U} : x = tF(x) + (1-t)p \text{ for some } t \in [0,1]\}.$$

Now $A \neq \varnothing$ since $p \in U$. In addition, the continuity of F implies that

A is closed. Notice that

$$A \cap \partial U = \varnothing,$$

therefore by Urysohn's lemma there exists a continuous $\mu : \overline{U} \to [0,1]$ with $\mu(A) = 1$ and $\mu(\partial U) = 0$. Let

$$N(x) := \begin{cases} \mu(x)F(x) + (1 - \mu(x))p, & x \in \overline{U}, \\ p, & x \in C \backslash \overline{U}. \end{cases}$$

Now it is immediate that $N : C \to C$ is a continuous, compact map. To see compactness use Mazur's theorem (see Exercise 4.12) together with

$$N(C) \subseteq \overline{\text{co}}(F(\overline{U}) \cup \{p\}).$$

Schauder's fixed point theorem guarantees the existence of $x \in C$ with $x = N(x)$. Notice that $x \in U$ since $p \in U$. Thus

$$x = \mu(x)F(x) + (1 - \mu(x))p.$$

As a result, $x \in A$ and therefore $\mu(x) = 1$. This implies that $x = F(x)$. □

Before we present generalisations of Theorem 5.1 we illustrate how Theorem 5.1 can be applied in practice. Consider the Fredholm integral equation

$$y(t) = h(t) + \int_0^1 k(t,s)g(s,y(s))\,ds \text{ for } t \in [0,1]. \tag{5.1}$$

A general existence principle will be established for (5.1) using Theorem 5.1. This principle can then be used to establish various existence results for the Fredholm integral equation.

Theorem 5.2 *Let $1 \le p \le \infty$ be a constant and q be such that $1/p + 1/q = 1$. Assume the following hold:*

$$h \in C[0,1]; \tag{5.2}$$

$$\begin{cases} g : [0,1] \times \mathbf{R} \to \mathbf{R} \textit{ is an } L^q\textit{-Carathéodory function;} \\ \textit{by this we mean} \\ \text{(i)} \;\; \textit{the map } t \mapsto g(t,y) \textit{ is measurable for all } y \in \mathbf{R}, \\ \text{(ii)} \;\; \textit{the map } y \mapsto g(t,y) \textit{ is continuous for almost all } t \in [0,1], \\ \text{(iii)} \;\; \textit{for any } r > 0, \;\; \textit{there exists } \mu_r \in L^q[0,1] \textit{ such that } |y| \le r \\ \qquad \textit{implies that } |g(t,y)| \le \mu_r(t) \textit{ for almost all } t \in [0,1]; \end{cases} \tag{5.3}$$

$$k_t(s) = k(t,s) \in L^p[0,1] \textit{ for each } t \in [0,1]; \tag{5.4}$$

and

(5.5) *the map* $t \mapsto k_t$ *is continuous from* $[0,1]$ *to* $L^p[0,1]$.

In addition, suppose there is a constant $M > 0$, *independent of* λ, *with*

$$|y|_0 = \sup_{t\in[0,1]} |y(t)| \neq M$$

for any solution $y \in C[0,1]$ *of*

$$(5.6)_\lambda \qquad y(t) = \lambda\left(h(t) + \int_0^1 k(t,s)g(s,y(s))\,ds\right),\ t \in [0,1],$$

for each $\lambda \in (0,1)$. *Then* (5.1) *has at least one solution* $y \in C[0,1]$.

Proof Define the operator F by

$$Fy(t) := h(t) + \int_0^1 k(t,s)g(s,y(s))\,ds \text{ for } t \in [0,1].$$

Notice that $F : C[0,1] \to C[0,1]$. To see this note that for any $y \in C[0,1]$, there exists $r > 0$ such that $|y|_0 \leq r$ and since g is L^q-Carathéodory, there exists $\mu_r \in L^q[0,1]$ with $|g(s,y)| \leq \mu_r(s)$, for almost every $s \in [0,1]$. Therefore for any t_1, $t_2 \in [0,1]$, we see that

$$(5.7) \quad \begin{aligned} |Fy(t_1) - Fy(t_2)| &\leq |h(t_1) - h(t_2)| \\ &\quad + \left(\int_0^1 |k_{t_1}(s) - k_{t_2}(s)|^p\,ds\right)^{\frac{1}{p}} \|\mu_r\|_q \\ &\to 0 \text{ as } t_1 \to t_2, \end{aligned}$$

(here $\|\mu_r\|_q = \|\mu_r\|_{L^q}$). Consequently, $F : C[0,1] \to C[0,1]$.

Now $(5.6)_\lambda$ is equivalent to the fixed point problem

$$y = \lambda F y.$$

We will apply Theorem 5.1 with

$$U := \{y \in C[0,1] : |y|_0 < M\} \text{ and } C = E = C[0,1].$$

First we show that $F : \overline{U} \to C[0,1]$ is continuous. Let $y_n \to y$ in $C[0,1]$ with $\{y_n\}_{n=1}^{\infty} \subseteq \overline{U}$. We are required to show that $Fy_n \to Fy$ in $C[0,1]$. There exists $\mu_M \in L^q[0,1]$ with $|y_n|_0 \leq M$, $|y|_0 \leq M$ and $|g(s,y_n(s))| \leq \mu_M(s)$, $|g(s,y(s))| \leq \mu_M(s)$ for almost all $s \in [0,1]$ (here $n \in \{1,2,\ldots\}$). By an argument similar to the one used to derive (5.7), we find that for any $\epsilon > 0$, there exists a $\delta > 0$ such that for $t_1, t_2 \in [0,1]$, with $|t_1 - t_2| < \delta$ we have

$$(5.8) \qquad |Fy_n(t_1) - Fy_n(t_2)| < \frac{\epsilon}{3} \text{ and } |Fy(t_1) - Fy(t_2)| < \frac{\epsilon}{3};$$

here $n \in \{1, 2, \ldots\}$. In addition, $Fy_n(t) \to Fy(t)$ pointwise on $[0,1]$, since the Lebesgue dominated convergence theorem implies that

$$(5.9) \qquad |Fy_n(t) - Fy(t)| \leq \left(\sup_{t\in[0,1]} \|k_t\|_p \right) \times \left(\int_0^1 |g(s, y_n(s)) - g(s, y(s))|^q \, ds \right)^{\frac{1}{q}} \to 0 \text{ as } n \to \infty.$$

Combining (5.8) and (5.9), and using the fact that $[0,1]$ is compact, yields a constant $N \geq 0$ such that

$$\text{for all } n \geq N, \quad |Fy_n(t) - Fy(t)| < \epsilon \text{ for all } t \in [0,1].$$

Consequently $Fy_n \to Fy$ in $C[0,1]$ and therefore $F : \overline{U} \to C[0,1]$ is continuous.

We now show that $F : \overline{U} \to C[0,1]$ is compact. There exists $\mu_M \in L^q[0,1]$ such that $|g(s, y(s))| \leq \mu_M(s)$ for almost every $s \in [0,1]$ and $y \in \overline{U}$. Since we are working in $C[0,1]$, we can use the Arzelà–Ascoli theorem to prove compactness. Clearly $F(\overline{U})$ is uniformly bounded since

$$|Fy|_0 \leq |h|_0 + \left(\sup_{t\in[0,1]} \|k_t\|_p \right) \|\mu_r\|_q \text{ for all } y \in \overline{U}.$$

Using an argument similar to the one used to obtain (5.7), one can see that $F(\overline{U})$ is equicontinuous. It follows from the Arzelà–Ascoli theorem that $F(\overline{U})$ is relatively compact and therefore $F : \overline{U} \to C[0,1]$ is a compact map.

We may now apply Theorem 5.1 (notice that possibility (A2) cannot occur) to deduce that F has a fixed point in $\overline{U}$, or equivalently, (5.1) has a solution in $\overline{U}$. □

We can now use our existence principle, Theorem 5.2, to obtain existence criteria for (5.1). To illustrate the ideas involved, we establish two general existence results for (5.1).

Theorem 5.3 *Let $1 \leq p \leq \infty$ be a constant and q be such that $1/p +$*

$1/q = 1$. *Assume that* (5.2)–(5.5) *are satisfied. In addition, suppose the following hold:*

$$(5.10)\quad \begin{cases} \text{there exists a continuous function } f : [0,\infty) \to [0,\infty), \\ \text{with } f(u) > 0 \text{ for } u > 0, \text{ and with } |u|_0 = \sup_{t\in[0,1]} |u(t)|, \\ \sup_{t\in[0,1]} \int_0^1 |k(t,s)g(s,u(s))|\,ds \le f(|u|_0) \text{ for any } u \in C[0,1] \end{cases}$$

and

$$(5.11)\qquad \sup_{c\in[0,\infty)} \left(\frac{c}{|h|_0 + f(c)} \right) > 1.$$

Then (5.1) *has at least one solution in* $C[0,1]$.

Remark 5.1 The supremum in (5.11) is allowed to be infinite.

Proof of Theorem 5.3 Let $M > 0$ satisfy

$$(5.12)\qquad \frac{M}{|h|_0 + f(M)} > 1.$$

Let $y \in C[0,1]$ be any solution of $(5.6)_\lambda$ for $0 < \lambda < 1$. Then for $t \in [0,1]$, we have that

$$|y(t)| \le |h|_0 + \int_0^1 |k(t,s)g(s,y(s))|\,ds \le |h|_0 + f(|y|_0),$$

and therefore

$$(5.13)\qquad \frac{|y|_0}{|h|_0 + f(|y|_0)} \le 1.$$

Suppose that $|y|_0 = M$. Then (5.13) implies that

$$\frac{M}{|h|_0 + f(M)} \le 1,$$

which contradicts (5.12). Thus any solution of $(5.6)_\lambda$ satisfies $|y|_0 \ne M$. Theorem 5.2 now guarantees that (5.1) has a solution $y \in C[0,1]$. □

Our next result arises when the nonlinearity g satisfies a particular monotonicity condition and the kernel k is of negative type.

Theorem 5.4 *Suppose that* (5.2)–(5.5) *hold with* $p = \infty$ *and* $q = 1$. *In addition, assume the following hold:*

$$(5.14)\qquad \begin{cases} \text{there exists a constant } R > 0 \text{ with } yg(t,y) \ge 0 \\ \text{for } |y| \ge R \text{ and almost every } t \in [0,1] \end{cases}$$

and

$$(5.15)\qquad \begin{cases} k \text{ is of negative type, that is, for any } u \in C[0,1], \\ \int_0^1 \int_0^1 k(t,s)g(s,y(s))g(t,y(t))\,ds\,dt \le 0. \end{cases}$$

Then (5.1) *has at least one solution* $y \in C[0,1]$.

Proof Let y be a solution of $(5.6)_\lambda$ for $0 < \lambda < 1$. Multiply $(5.6)_\lambda$ by $g(t, y(t))$ and integrate from 0 to 1 to obtain

$$(5.16)\qquad \int_0^1 y(t)g(t,y(t))\,dt \le \lambda \int_0^1 h(t)g(t,y(t))\,dt.$$

Let

$$R_0 := \max\{R, |h|_0 + 1\}.$$

Note that $yg(t,y) \ge 0$ for $|y| \ge R_0$ and almost every $t \in [0,1]$. In addition, let

$$I_0 := \{t \in [0,1] : |y(t)| \ge R_0\} \text{ and } J_0 := \{t \in [0,1] : |y(t)| < R_0\}.$$

Notice that

$$(5.17)\qquad \begin{aligned} \int_{I_0} y(t)g(t,y(t))\,dt &= \int_{I_0} |y(t)g(t,y(t))|\,dt \\ &\ge R_0 \int_{I_0} |g(t,y(t))|\,dt. \end{aligned}$$

Put (5.16) into (5.17) to obtain

$$(5.18)\qquad \begin{aligned} R_0 \int_{I_0} |g(t,y(t))|\,dt &\le \int_0^1 |h(t)g(t,y(t))|\,dt \\ &\quad + \int_{J_0} |y(t)g(t,y(t))|\,dt. \end{aligned}$$

Since g is an L^1-Carathéodory function, there exists $\mu_{R_0} \in L^1[0,1]$ with $|g(t,u)| \le \mu_{R_0}(t)$ for almost every $t \in [0,1]$, and all $|u| \le R_0$. Consequently,

$$(5.19)\qquad \begin{aligned} &\int_{J_0} |h(t)g(t,y(t))|\,dt + \int_{J_0} |y(t)g(t,y(t))|\,dt \\ &\le \int_{J_0} |h(t)|\mu_{R_0}(t)\,dt + R_0 \int_0^1 \mu_{R_0}(t)\,dt \\ &\le (R_0 + |h|_0)\|\mu_{R_0}\|_1. \end{aligned}$$

Putting (5.19) into (5.18), we obtain

$$\begin{aligned} R_0 \int_{I_0} |g(t,y(t))|\,dt &\leq \int_{I_0} |h(t)g(t,y(t))|\,dt + (R_0 + |h|_0)\|\mu_{R_0}\|_1 \\ &\leq |h|_0 \int_{I_0} |g(t,y(t))|\,dt + (R_0 + |h|_0)\|\mu_{R_0}\|_1. \end{aligned}$$

This immediately yields

$$\int_{I_0} |g(t,y(t))|\,dt \leq \left(\frac{R_0 + |h|_0}{R_0 - |h|_0}\right) \|\mu_{R_0}\|_1. \tag{5.20}$$

Returning to $(5.6)_\lambda$ and using (5.20), we have for $t \in [0,1]$ that

$$\begin{aligned} |y(t)| &\leq |h|_0 + \int_{I_0} |k(t,s)g(s,y(s))|\,ds + \int_{J_0} |k(t,s)g(s,y(s))|\,ds \\ &\leq |h|_0 + \sup_{t\in[0,1]} \|k_t\|_{L^\infty} \left[\int_{I_0} |g(s,y(s))|\,ds + \int_{J_0} |g(s,y(s))|\,ds\right] \\ &\leq |h|_0 + \sup_{t\in[0,1]} \|k_t\|_{L^\infty} \left[\left(\frac{R_0 + |h|_0}{R_0 - |h|_0}\right) \|\mu_{R_0}\|_1 + \|\mu_{R_0}\|_1\right] \\ &\equiv M_0, \end{aligned}$$

that is, $|y|_0 \leq M_0$. Now apply Theorem 5.2 with $M > M_0$. □

We now extend Theorem 5.1 to Mönch type maps.

Theorem 5.5 *Let E be a Banach space, C a closed, convex subset of E, U an open subset of C and $p \in U$. Suppose that $F : \overline{U} \to C$ is a continuous map which satisfies Mönch's condition (that is, if $D \subseteq \overline{U}$ is countable and $D \subseteq \overline{\text{co}}(\{p\} \cup F(D))$, then $\overline{D}$ is compact) and assume that*

$$x \neq tF(x) + (1-t)p \text{ for } x \in \partial U \text{ and } t \in (0,1) \tag{5.21}$$

holds. Then F has a fixed point in $\overline{U}$.

Proof Assume that F has no fixed point on ∂U (otherwise we are finished). Then

$$x \neq tF(x) + (1-t)p \text{ for } x \in \partial U \text{ and } t \in [0,1]. \tag{5.22}$$

Consider

$$A := \{x \in \overline{U} : x = tF(x) + (1-t)p \text{ for some } t \in [0,1]\}.$$

Now $A \neq \varnothing$ is closed and $A \cap \partial U = \varnothing$. Thus there exists a continuous map $\mu : \overline{U} \to [0,1]$ with $\mu(A) = 1$ and $\mu(\partial U) = 0$. Let $N : C \to C$ be defined by

$$N(x) := \begin{cases} \mu(x)F(x) + (1-\mu(x))p, & x \in \overline{U}, \\ \{p\}, & x \in C \backslash \overline{U}. \end{cases}$$

Now $N : C \to C$ is continuous. In addition, N satisfies Mönch's condition. To see this let $D \subseteq C$ be countable and

$$D \subseteq \overline{\text{co}}(\{p\} \cup N(D)).$$

We must show that $\overline{D}$ is compact. Notice first that

$$N(D) \subset \text{co}(F(D \cap \overline{U}) \cup \{p\}).$$

Also, since $\text{co}(F(D \cap \overline{U}) \cup \{p\})$ is convex and

$$\{p\} \cup \text{co}(F(D \cap \overline{U}) \cup \{p\}) = \text{co}(F(D \cap \overline{U}) \cup \{p\})$$

we have

$$\begin{aligned} D &\subseteq \overline{\text{co}}(\{p\} \cup \text{co}(F(D \cap \overline{U}) \cup \{p\})) \\ &= \overline{\text{co}}(\text{co}(F(D \cap \overline{U}) \cup \{p\})) = \overline{\text{co}}(F(D \cap \overline{U}) \cup \{p\}). \end{aligned}$$

Notice also that $D \cap \overline{U}$ is countable and

$$D \cap \overline{U} \subseteq \overline{\text{co}}(F(D \cap \overline{U}) \cup \{p\}).$$

Since $F : \overline{U} \to E$ satisfies Mönch's condition, we have that $\overline{D \cap \overline{U}}$ is compact. Thus since F is continuous, $F(\overline{D \cap \overline{U}})$ is compact and Mazur's theorem (Exercise 4.12) guarantees that

$$\overline{\text{co}}(F(\overline{D \cap \overline{U}}) \cup \{p\})$$

is compact. Also, since

$$D \subseteq \overline{\text{co}}(F(\overline{D \cap \overline{U}}) \cup \{p\}),$$

we have that $\overline{D}$ is compact. Consequently, $N : C \to C$ is continuous and satisfies Mönch's condition. Mönch's fixed point theorem guarantees the existence of $x \in C$ with $x = N(x)$. Notice that $x \in U$ since $p \in U$. Thus

$$x = \mu(x)F(x) + (1-\mu(x))p$$

and therefore $x \in A$. Consequently, $\mu(x) = 1$ which implies that $x = F(x)$. $\square$

Let E be a Banach space and Ω_E be the bounded subsets of E. The Kuratowskii measure of noncompactness (see Chapter 4) is the map $\alpha : \Omega_E \to [0, \infty)$ defined by (here $A \in \Omega_E$)

$$\alpha(A) := \inf \left\{ \epsilon > 0 : A \subseteq \bigcup_{i=1}^{n} A_i \text{ and } \operatorname{diam}(A_i) \leq \epsilon \text{ for } i = 1, \ldots, n \right\}.$$

Definition 5.1 A map $F : X \subseteq E \to E$ is said to be *condensing* if $\alpha(F(Y)) < \alpha(Y)$ for all bounded sets $Y \subseteq X$ with $\alpha(Y) \neq 0$.

Definition 5.2 A map $F : X \subseteq E \to E$ is said to be *k-set contractive* (here $k \geq 0$ is a constant) if $\alpha(F(Y)) \leq k\alpha(Y)$ for all bounded sets $Y \subseteq X$.

Definition 5.3 A map $F : X \subseteq E \to E$ is said to be *completely continuous* if $F(Y)$ is relatively compact for all bounded sets $Y \subseteq X$.

Theorem 5.6 *Let E be a Banach space, C a closed, convex subset of E, U an open subset of C and $p \in U$. Suppose that $F : \overline{U} \to C$ is a continuous, condensing map with $F(\overline{U})$ a bounded set in C and assume that*

(5.23) $\qquad x \neq tF(x) + (1-t)p$ *for* $x \in \partial U$ *and* $t \in (0, 1)$

holds. Then F has a fixed point in $\overline{U}$.

Proof We will apply Theorem 5.5. Let $D \subseteq \overline{U}$ be countable and $D \subseteq \overline{\text{co}}(\{p\} \cup F(D))$. If $\alpha(D) \neq 0$ then

$$\alpha(D) \leq \alpha(\overline{\text{co}}(\{p\} \cup F(D))) = \alpha(F(D)) < \alpha(D)$$

– a contradiction. Thus $\alpha(D) = 0$ and therefore $\overline{D}$ is compact. □

Theorem 5.7 *Let E be a Banach space, C a closed, convex subset of E, U an open subset of E and $p \in U$. Suppose that $F : \overline{U} \to C$ is a continuous, k-set contractive (here $0 \leq k < 1$ is a constant) map with $F(\overline{U})$ a bounded set in C. Then either*

(A1) *F has a fixed point in $\overline{U}$, or*
(A2) *there exist $u \in \partial U$ and $\lambda \in (0, 1)$ with $u = \lambda F(u) + (1-\lambda)p$.*

Proof Theorem 5.7 is a direct consequence of Theorem 5.6. □

A special case of Theorem 5.6 is the following result of Krasnoselskii type.

Theorem 5.8 *Let $(E, \|\cdot\|)$ be a Banach space, C a closed, convex subset of E, U an open subset of C and $p \in U$. Suppose that $F : \overline{U} \to C$ is given by $F := F_1 + F_2$ and $F(\overline{U})$ is a bounded set in C. In addition, assume that*

$$F_1 : \overline{U} \to C \text{ is continuous and completely continuous}$$

and

$$F_2 : \overline{U} \to C \text{ is a nonlinear contraction}$$

(that is, there exists a continuous, nondecreasing function $\phi : [0,\infty) \to [0,\infty)$ with $\phi(z) < z$ for $z > 0$, such that $\|F_2(x) - F_2(y)\| \le \phi(\|x - y\|)$ for all x, $y \in \overline{U}$). Then either

(A1) *F has a fixed point in $\overline{U}$, or*

(A2) *there are a $u \in \partial U$ and $\lambda \in (0,1)$ with $u = \lambda F(u) + (1-\lambda)p$.*

Proof We will apply Theorem 5.6 to obtain the desired result. Let Ω be a bounded subset of $\overline{U}$. Then we have

$$\alpha(F(\Omega)) \le \alpha(F_1(\Omega)) + \alpha(F_2(\Omega)) = \alpha(F_2(\Omega)), \tag{5.24}$$

since $F_1 : \overline{U} \to E$ is completely continuous. We now claim that

$$\alpha(F_2(\Omega)) \le \phi(\alpha(\Omega)) \tag{5.25}$$

holds. If this is true, then (5.24) and (5.25) yield

$$\alpha(F(\Omega)) \le \phi(\alpha(\Omega))$$

and consequently, $F : \overline{U} \to C$ is a condensing map. To show that (5.25) holds, let $\epsilon > 0$. Then there exist $\Omega_1, \dots, \Omega_n$ with

$$\Omega \subseteq \bigcup_{i=1}^{n} \Omega_i \text{ and } \operatorname{diam}(\Omega_i) \le \alpha(\Omega) + \epsilon.$$

Now

$$F_2(\Omega) \subseteq \bigcup_{i=1}^{n} F_2(\Omega_i) \equiv \bigcup_{i=1}^{n} Y_i.$$

If w_0, $w_1 \in Y_i$ for some i, then there exist x_0, $x_1 \in \Omega_i$ with $F_2(x_0) = w_0$ and $F_2(x_1) = w_1$. Since ϕ is nondecreasing we obtain

$$\|F_2(x_0) - F_2(x_1)\| \le \phi(\|x_0 - x_1\|) \le \phi(\alpha(\Omega) + \epsilon).$$

Consequently,

$$\text{diam}(Y_i) \leq \phi(\alpha(\Omega) + \epsilon) \text{ for } i = 1, \ldots, n$$

and as a result

$$\alpha(F_2(\Omega)) \leq \phi(\alpha(\Omega) + \epsilon).$$

Since $\epsilon > 0$ is arbitrary, (5.25) follows. □

Remark 5.2 Notice a special case of Theorem 5.8 is when $F_1 : \overline{U} \to E$ is a contraction (with Lipschitz constant $L < 1$). To see this, take $\phi(z) = Lz$. In this particular case notice that $F : \overline{U} \to E$ is L-set contractive.

Our next result again concerns the sum of two operators, one being compact and the other nonexpansive.

Theorem 5.9 *Let $E = (E, \|\cdot\|) = C$ be a uniformly convex Banach space and U a bounded, open, convex subset of E with $p \in U$. Suppose that $F : \overline{U} \to E$ is given by $F := F_1 + F_2$ where*

$$F_1 : \overline{U} \to E \textit{ is strongly continuous}$$

and

$$F_2 : \overline{U} \to E \textit{ is a nonexpansive map.}$$

Then either

(A1) *F has a fixed point in $\overline{U}$, or*
(A2) *there are a $u \in \partial U$ and $\lambda \in (0, 1)$ with $u = \lambda F(u) + (1 - \lambda)p$.*

Remark 5.3 $F_1 : \overline{U} \to E$ is said to be *strongly continuous* if $\{x_n\}_{n=1}^{\infty} \subseteq \overline{U}$, $x_n \rightharpoonup x$ imply that $F_1(x_n) \to F_1(x)$; here $\rightharpoonup$ denotes weak convergence.

Remark 5.4 Notice that $F_1 : \overline{U} \to E$ strongly continuous in Theorem 5.9 implies that $F_1 : \overline{U} \to E$ is continuous and compact. We need only show compactness. Notice first that since U is convex and E is reflexive, then $\overline{U}$ is weakly closed. In addition since U is bounded, we have that $\overline{U}$ is weakly compact. Now let $\{y_n\}_{n=1}^{\infty} \subseteq F_1(\overline{U})$. Then for each $n \in \{1, 2, \ldots\}$, there exists $x_n \in \overline{U}$ with $y_n = F_1(x_n)$. Since $\overline{U}$ is weakly compact, there exists a subsequence S of $\{1, 2, \ldots\}$ with $x_n \rightharpoonup x \in \overline{U}$ as $n \to \infty$ in S. Due to the strong continuity of F_1 we have that $y_n = F_1(x_n) \to F(x_0)$ as $n \to \infty$ in S.

Proof of Theorem 5.9 Assume that (A2) does not hold. Without loss of generality, assume that $p = 0$. Consider for each $n \in \{2, 3, \ldots\}$, the mapping

$$G_n := \left(1 - \frac{1}{n}\right) F : \overline{U} \to E.$$

Notice that $(1 - 1/n)F_2 : \overline{U} \to E$ is a contraction, and $(1 - 1/n)F_1 : \overline{U} \to E$ is continuous and compact. We want to apply Theorem 5.8 to G_n. If there exist $\lambda \in (0, 1)$ and $u \in \partial U$ with $u = \lambda G_n(u)$, then

$$u = \lambda \left(1 - \frac{1}{n}\right) F(u) = \eta F(u) \text{ where } 0 < \eta = \lambda \left(1 - \frac{1}{n}\right) < 1,$$

which is a contradiction since (A2) was assumed not to hold. Consequently, for each $n \in \{2, 3, \ldots\}$, we have from Theorem 5.8 that G_n has a fixed point $u_n \in \overline{U}$.

Since E is a reflexive Banach space (see Exercise 5.5), then any norm bounded sequence in E has a weakly convergent subsequence. Therefore since $\overline{U}$ is bounded, there exist a subsequence S of integers and a $u \in \overline{U}$ (notice that $\overline{U}$ is strongly closed and convex, therefore weakly closed) with

$$u_n \rightharpoonup u \text{ as } n \to \infty \text{ in } S.$$

In addition,

$$\begin{aligned} \|(I - F_2)(u_n) - F_1(u)\| &= \left\| -\frac{1}{n} F_2(u_n) + F_1(u_n) - F_1(u) - \frac{1}{n} F_1(u_n) \right\| \\ &\leq \frac{1}{n} \{\|u_n\| + \|F_2(0)\| + \|F_1(u_n)\|\} \\ &\quad + \|F_1(u_n) - F_1(u)\|, \end{aligned}$$

therefore since $F_1 : \overline{U} \to E$ is strongly continuous, we have that

$$(I - F_2)(u_n) \to F_1(u) \text{ as } n \to \infty \text{ in } S.$$

The demiclosedness of $I - F_2$ (see Exercise 2.8) implies that $F_1(u) = (I - F_2)(u)$. □

All the results so far in Chapter 5 concern maps $F : X \subseteq E \to E$, when the interior of X is nonempty. We now present some results when the interior of X may be empty.

Theorem 5.10 *Let E be a Hilbert space, Q a closed, convex subset of E and $0 \in Q$. Suppose that $F : Q \to E$ is a continuous, k-set contractive* *(here $0 \leq k < 1$ is a constant) map with $F(Q)$ a bounded set in E. Assume also that the following condition holds:*

$$(5.26) \quad \begin{cases} \text{if } \{(x_j, \lambda_j)\}_{j=1}^{\infty} \text{ is a sequence in } \partial Q \times [0,1] \\ \text{converging to } (x, \lambda) \text{ with } x = \lambda F(x) \text{ and } 0 \leq \lambda < 1, \\ \text{then } \lambda_j F(x_j) \in Q \text{ for } j \text{ sufficiently large.} \end{cases}$$

Then F has a fixed point in Q.

Proof Define $r : E \to Q$ by

$$r(x) := P_Q(x),$$

that is, the nearest point projection on Q. We know (Exercise 2.5) that r is continuous and nonexpansive. Consider

$$B := \{x \in E : x = Fr(x)\}.$$

We first show that B is nonempty. To see this we consider $rF : Q \to Q$. Since F is k-set contractive and r is nonexpansive, we have that rF is k-set contractive. Also, $r(F(Q))$ is bounded since $F(Q)$ is bounded and r is nonexpansive, that is, for $u \in Q$ we have that

$$\|r(F(u))\| \leq \|r(F(0))\| + \|F(0)\| + \|F(u)\|.$$

Mönch's fixed point theorem (or Exercise 4.5) guarantees that rF has a fixed point. This immediately implies that Fr has a fixed point and therefore $B \neq \varnothing$. The continuity of F and r implies that B is closed. Next we show that B is compact. To see this notice that $B \subseteq F(r(B))$. Since r is nonexpansive and F is k-set contractive, we see that

$$\alpha(B) \leq \alpha(F(r(B))) \leq k\alpha(r(B)) \leq k\alpha(B).$$

Thus $\alpha(B) = 0$ and therefore B is compact.

We now show that $B \cap Q \neq \varnothing$. Suppose this is not true, that is, suppose that $B \cap Q = \varnothing$. Then since B is compact and Q is closed, there exists $\delta > 0$ with $\text{dist}(B, Q) > \delta$. Choose $N \in \{1, 2, \ldots\}$ with $1 < \delta N$. Define

$$U_i := \left\{x \in E : d(x, Q) < \frac{1}{i}\right\} \text{ for } i \in \{N, N+1, \ldots\};$$

here d denotes the metric induced by the norm. Fix $i \in \{N, N+1, \ldots\}$. Since $\text{dist}(B, Q) > \delta$, we have that $B \cap \overline{U_i} = \varnothing$. Notice that U_i is open and also $Fr : \overline{U_i} \to E$ is k-set contractive. Theorem 5.7 guarantees, since $B \cap \overline{U_i} \neq \varnothing$, that there exists $(y_i, \lambda_i) \in \partial U_i \times (0, 1)$ with $y_i = \lambda_i Fr(y_i)$.

Thus for each $j \in \{N, N+1, \dots\}$ there exists $(y_j, \lambda_j) \in \partial U_j \times (0,1)$ with $y_j = \lambda_j Fr(y_j)$. Consequently,

$$\lambda_j Fr(y_j) \notin Q \text{ for } j \in \{N, N+1, \dots\}. \tag{5.27}$$

We now look at

$$D := \{x \in E : x = \lambda Fr(x) \text{ for some } \lambda \in [0,1]\}.$$

Notice that D is closed and in fact compact. To see this note that

$$D \subseteq \overline{\text{co}}(F(r(D)) \cup \{0\}),$$

and therefore

$$\begin{aligned} \alpha(D) &\leq \alpha(\overline{\text{co}}(Fr(D) \cup \{0\})) \\ &= \alpha(Fr(D)) \leq k\alpha(r(D)) \leq k\alpha(D). \end{aligned}$$

Hence $\alpha(D) = 0$ and therefore D is compact (therefore, sequentially compact). This, together with

$$d(y_j, Q) = \frac{1}{j} \text{ and } |\lambda_j| \leq 1 \text{ for } j \in \{N, N+1, \dots\},$$

implies that we may assume without loss of generality that

$$\lambda_j \to \lambda^\star \in [0,1] \text{ and } y_j \to y^\star \in \partial Q.$$

Also we have that

$$y_j = \lambda_j Fr(y_j) \to \lambda^\star Fr(y^\star),$$

and therefore $y^\star = \lambda^\star Fr(y^\star)$. Now $\lambda^\star \neq 1$ since $B \cap Q = \varnothing$. Hence $0 \leq \lambda^\star < 1$. However, (5.26) with

$$x_j = r(y_j) \in \partial Q \text{ and } x = y^\star = r(y^\star),$$

implies that $\lambda_j Fr(y_j) \in Q$ for j sufficiently large. This contradicts (5.27). Thus $B \cap Q = \varnothing$ and therefore there exists $x \in Q$ with $x = Fr(x) = F(x)$. □

Remark 5.5 If $F(\partial Q) \subseteq Q$ in Theorem 5.10, then clearly (5.26) holds.

Remark 5.6 In Theorem 5.10, if $0 \in \text{int}(Q)$, where $\text{int}\, Q$ denotes the interior of Q, then the proof would be a lot simpler. One would just need to show that condition (A2) in Theorem 5.7 is not satisfied.

Remark 5.7 In Theorem 5.10, $F : Q \to E$ being k-set contractive, $0 \le k < 1$, could be replaced by $F : Q \to E$ being 1-set contractive and condensing.

Remark 5.8 If the map $F : Q \to E$ is a continuous, compact map, then the Hilbert space E could be replaced by a Banach space. The proof follows the same reasoning as in Theorem 5.10, except in this case $r : E \to Q$ is a continuous retraction which exists by Dugundji's extension theorem (Exercise 4.8). Notice that r is *not* necessarily nonexpansive; however, we need only the continuity of r in the proof of Theorem 5.10 if $F : Q \to E$ is compact.

Our final theorem in this chapter is a result of Reinermann type (see Exercise 5.9).

Theorem 5.11 *Let E be a Hilbert space, Q a closed, convex, bounded subset of E and $0 \in Q$. Suppose that $F : Q \to E$ is given by $F := F_1 + F_2$, where*

$$F_1 : Q \to E \text{ is strongly continuous}$$

and

$$F_2 : Q \to E \text{ is nonexpansive.}$$

In addition, assume that (5.26) *holds. Then F has a fixed point in Q.*

Proof For each $n \in \{2, 3, \ldots\}$, consider the mapping

$$G_n := \left(1 - \frac{1}{n}\right) F : Q \to E.$$

Notice that $(1 - 1/n)F_1 : Q \to E$ is continuous and compact, while $F_2 : Q \to E$ is a contraction. We wish to apply Theorem 5.10 to G_n. Let $\{(x_j, \lambda_j)\}_{j=1}^{\infty}$ be a sequence in $\partial Q \times [0, 1]$ converging to (x, λ) with $x = \lambda G_n(x)$ and $0 \le \lambda < 1$. Then

$$\lambda_j G_n(x_j) = \lambda_j \left(1 - \frac{1}{n}\right) F(x_j) \equiv \mu_j F(x_j) \in Q \text{ for } j \text{ sufficiently large,}$$

since F satisfies (5.26); note that $\mu_j = \lambda_j(1 - 1/n)$ is a sequence in $[0, 1)$ with

$$\mu_j \to \lambda(1 - 1/n) \equiv \mu \text{ as } j \to \infty, \quad 0 \le \mu < 1,$$

and

$$x = \lambda G_n(x) = \lambda \left(1 - \frac{1}{n}\right) F(x) = \mu F(x).$$

Apply Theorem 5.10 to G_n to deduce that G_n has a fixed point $u_n \in Q$. Now since Q is bounded, there exist a subsequence S of integers and a $u \in Q$ with

$$u_n \rightharpoonup u \text{ as } n \to \infty \text{ in } S.$$

Essentially the same reasoning as in Theorem 5.9 now yields

$$(I - F_2)(u_n) \to F_1(u) \text{ as } n \to \infty \text{ in } S.$$

The demiclosedness of $I - F_2$ implies that $F_1(u) = (I - F_2)(u)$. $\square$

Notes The nonlinear alternatives of Leray–Schauder type (that is, Theorem 5.1 and Theorem 5.5) were adapted from Banas and Goebel [13], Granas [85], O'Regan [141] and Precup [154]. Theorems 5.2–5.3 can be found in O'Regan and Meehan [144], and Brezis and Browder [26]. Theorem 5.10 and Theorem 5.11 were adapted from Furi and Pera [72] and O'Regan [139, 141].

Exercises

5.1 Let E be a Banach space, C a closed, convex subset of E and $0 \in C$. Let $F : C \to C$ be a continuous and completely continuous map. Define

$$E(F) := \{x \in C : x = \lambda F(x) \text{ for some } \lambda \in (0,1)\}.$$

Show that either $E(F)$ is unbounded or F has a fixed point.

5.2 Let E be a normed linear space. A map $F : E \to E$ is called *quasibounded* whenever

$$|F|_q = \inf_{\rho>0} \sup_{\|x\|\geq\rho} \frac{\|F(x)\|}{\|x\|} = \lim \sup_{\|x\|\to\infty} \frac{\|F(x)\|}{\|x\|} < \infty.$$

($|F|_q$ is called the *quasinorm* of F.)

(a) Show that every bounded, linear operator is quasibounded.

(b) Let $F : E \to E$ be a quasibounded, continuous and completely continuous map. Show for each $|\lambda| < 1/|F|_q$ (and for all real λ when $|F|_q = 0$) that λF has at least one fixed point.

5.3 Let E be a Banach space, C a closed, convex subset of E, U an open subset of C with $0 \in U$. Suppose that $F : \overline{U} \to C$ is a continuous, k-set contractive (here $0 \leq k < 1$) map with

$$\|F(x)\| \leq \sqrt{\|x\|^2 + \|x - F(x)\|^2} \text{ for } x \in \partial U.$$

Show that F has a fixed point in $\overline{U}$.

5.4 Let E be a Banach space, C a closed, convex subset of E, U an open subset of C and $p \in U$. Suppose that $F : \overline{U} \to C$ is a 1-set contractive map with $F(\overline{U})$ a bounded set in C and $(I - F)(\overline{U})$ closed. Show that either

(A1) F has a fixed point in $\overline{U}$, or
(A2) there exist a $u \in \partial U$ and $\lambda \in (0,1)$ with $u = \lambda F(u) + (1-\lambda)p$.

5.5 Show that every uniformly convex Banach space is reflexive.

5.6 Let E be a Banach space, Q a closed, convex subset of E and $0 \in Q$. Suppose that $F : Q \to E$ is a continuous, compact map and that (5.26) holds. Show that F has a fixed point in Q.

5.7 Let E be a Hilbert space, Q a closed, convex subset of E and $0 \in Q$. Suppose that $F : Q \to E$ is a 1-set contractive map with $F(Q)$ a bounded set in C. Also assume that $(I - F)(Q)$ is a closed and (5.26) holds. Show that F has a fixed point in Q.

5.8 Let Q be a subset of a Banach space $E = (E, \|\cdot\|)$. Suppose that $F, F_n : Q \to E$ (here $n \in \{1, 2, \ldots\}$) with $(I - F)(Q)$ closed and

$$\lim_{n \to \infty} \sup_{x \in Q} \|F(x) - F_n(x)\| = 0.$$

If for each $n \in \{1, 2, \ldots\}$, F_n has a fixed point in Q, show that F has a fixed point in Q.

5.9 Let E be a uniformly convex Banach space, Q a closed, convex, bounded subset of E with $0 \in Q$. Suppose that $F : Q \to Q$ is given by $F := F_1 + F_2$, where $F_1 : Q \to E$ is strongly continuous and $F_2 : Q \to E$ is nonexpansive. Show that F has a fixed point in Q.

5.10 In Theorem 5.1 show that the Banach space E can be replaced by a normed linear space E. Also show that the condition that C is closed can be removed.

6
Continuation Principles for Condensing Maps

There are three main approaches in the literature to presenting continuation principles for condensing maps. The first uses degree theory (see Chapter 12), the second is the essential map approach of Granas and the third is the 0-epi map approach of Furi, Martelli and Vignoli. In this chapter we will present the second approach. We remark that the third approach will be presented in Chapter 8 in a more general setting. In particular in this chapter we will show that the property of having a fixed point (or more generally, being essential) is invariant by homotopy for compact (or more generally, condensing) maps.

We begin with compact maps. Throughout this chapter, E will be a Banach space, C will be a closed, convex subset of E and U will be an open subset of C.

Definition 6.1 We let $K(\overline{U}, C)$ denote the set of all continuous, compact maps $F : \overline{U} \to C$; here $\overline{U}$ denotes the closure of U in C.

Definition 6.2 We let $K_{\partial U}(\overline{U}, C)$ denote the set of maps $F \in K(\overline{U}, C)$ with $x \neq F(x)$ for $x \in \partial U$.

Definition 6.3 A map $F \in K_{\partial U}(\overline{U}, C)$ is *essential* in $K_{\partial U}(\overline{U}, C)$ if for every map $G \in K_{\partial U}(\overline{U}, C)$ with $G|_{\partial U} = F|_{\partial U}$, there exists $x \in U$ with $x = G(x)$. Otherwise F is *inessential* in $K_{\partial U}(\overline{U}, C)$, that is, there exists a fixed point free $G \in K_{\partial U}(\overline{U}, C)$ with $G|_{\partial U} = F|_{\partial U}$.

Remark 6.1 If $F \in K_{\partial U}(\overline{U}, C)$ is essential then there exists $x \in U$ with $x = F(x)$.

Definition 6.4 Two maps F and G belonging to $K_{\partial U}(\overline{U}, C)$ are *homotopic* in $K_{\partial U}(\overline{U}, C)$, written

$$F \simeq G \text{ in } K_{\partial U}(\overline{U}, C),$$

if there exists a continuous, compact mapping $H : \overline{U} \times [0, 1] \to C$ such that $H_t(\cdot) := H(\cdot, t) : \overline{U} \to C$ belongs to $K_{\partial U}(\overline{U}, C)$ for each $t \in [0, 1]$, with $H_0 = F$ and $H_1 = G$.

Theorem 6.1 *Let E be a Banach space, C a closed, convex subset of E, U an open subset of C and F, $G \in K_{\partial U}(\overline{U}, C)$. Suppose that*

$$x \neq tG(x) + (1 - t)F(x) \textit{ for each } (x, t) \in \partial U \times [0, 1].$$

Then $F \simeq G$ in $K_{\partial U}(\overline{U}, C)$.

Proof Let

$$H(x, t) := tG(x) + (1 - t)F(x) \text{ for } (x, t) \in \overline{U} \times [0, 1].$$

We first show that $H : \overline{U} \times [0, 1] \to C$ is a continuous, compact map. We need only show compactness. Let $\{(x_n, t_n)\}_{n=1}^{\infty}$ be any sequence in $\overline{U} \times [0, 1]$. Without loss of generality, assume that there exists $t \in [0, 1]$ with $\lim_{n \to \infty} t_n = t$. Since $F : \overline{U} \to C$ and $G : \overline{U} \to C$ are compact maps, there exist a subsequence S of integers and $u, v \in C$ with

$$F(x_n) \to v \text{ and } G(x_n) \to u \text{ as } n \to \infty \text{ in } S.$$

Due to the convexity of C we have that

$$\begin{aligned} H(x_n, t_n) &= t_n G(x_n) + (1 - t_n)F(x_n) \\ &\to tu + (1 - t)v \in C \text{ as } n \to \infty \text{ in } S. \end{aligned}$$

As a result, $H : \overline{U} \times [0, 1] \to C$ is a compact map. In addition, since

$$x \neq tG(x) + (1 - t)F(x) \text{ for } (x, t) \in \partial U \times [0, 1],$$

we have that $H_t \in K_{\partial U}(\overline{U}, C)$ for each $t \in [0, 1]$. Finally, $H_0 = F$ and $H_1 = G$, therefore $F \simeq G$ in $K_{\partial U}(\overline{U}, C)$. □

We next characterise inessential maps in $K_{\partial U}(\overline{U}, C)$ in terms of homotopy.

Theorem 6.2 *Let E be a Banach space, C a closed, convex subset of E, U an open subset of C and $F \in K_{\partial U}(\overline{U}, C)$. Then the following conditions are equivalent:*

(i) *F is inessential in $K_{\partial U}(\overline{U}, C)$,*
(ii) *there exists a fixed point free map $G \in K_{\partial U}(\overline{U}, C)$ with $F \simeq G$ in $K_{\partial U}(\overline{U}, C)$.*

Proof We first show that (i) implies (ii). Let $G \in K_{\partial U}(\overline{U}, C)$ be a fixed point free map with $F|_{\partial U} = G|_{\partial U}$. Suppose that there exist an $x \in \partial U$ and a $t \in [0,1]$ with

$$x = tG(x) + (1-t)F(x).$$

Then since $F|_{\partial U} = G|_{\partial U}$, we have that $x = G(x)$ – a contradiction since $G \in K_{\partial U}(\overline{U}, C)$. As a result,

$$x \neq tG(x) + (1-t)F(x) \text{ for each } (x,t) \in \partial U \times [0,1].$$

Thus Theorem 6.1 implies that $F \simeq G$ in $K_{\partial U}(\overline{U}, C)$.

We next show that (ii) implies (i). Let $H : \overline{U} \times [0,1] \to C$ be a continuous, compact map with $H_t \in K_{\partial U}(\overline{U}, C)$ for each $t \in [0,1]$, with $H_0 = G$ and $H_1 = F$. Consider

$$B := \{x \in \overline{U} : x = H(x,t) \text{ for some } t \in [0,1]\}.$$

If $B = \varnothing$ then for each $t \in [0,1]$ we have that H_t has no fixed points in $\overline{U}$, therefore in particular, F has no fixed points in $\overline{U}$. Thus F is inessential in $K_{\partial U}(\overline{U}, C)$. Therefore it remains to consider the case when $B \neq \varnothing$. The continuity of H implies that B is closed in $\overline{U}$. Since $B \cap \partial U = \varnothing$, there exists a continuous function

$$\mu : \overline{U} \to [0,1] \text{ with } \mu(\partial U) = 1 \text{ and } \mu(B) = 0.$$

Define $J : \overline{U} \to C$ by

$$J(x) := H(x, \mu(x)).$$

It is easy to see that $J : \overline{U} \to C$ is a continuous, compact map. In addition, $J|_{\partial U} = F|_{\partial U}$ since if $x \in \partial U$ then

$$J(x) = H(x,1) = F(x).$$

Finally, $x \neq J(x)$ for $x \in \overline{U}$ since $x = J(x)$ for some $x \in \overline{U}$ means that $x \in B$ and therefore $\mu(x) = 0$ (that is, $x = G(x)$) – a contradiction. Consequently, $J \in K_{\partial U}(\overline{U}, C)$ with $x \neq J(x)$ for $x \in \overline{U}$ and $J|_{\partial U} = F|_{\partial U}$. As a result, F is inessential in $K_{\partial U}(\overline{U}, C)$. □

Our next result shows that the property of being essential is invariant by homotopy for compact maps.

Theorem 6.3 *Let E be a Banach space, C a closed, convex subset of E and U an open subset of C. Suppose that F and G are two maps in $K_{\partial U}(\overline{U},C)$ with $F \simeq G$ in $K_{\partial U}(\overline{U},C)$. Then F is essential in $K_{\partial U}(\overline{U},C)$ if and only if G is essential in $K_{\partial U}(\overline{U},C)$.*

Proof If F is inessential in $K_{\partial U}(\overline{U},C)$ then Theorem 6.2 guarantees that there exists a fixed point free $T \in K_{\partial U}(\overline{U},C)$ with $F \simeq T$ in $K_{\partial U}(\overline{U},C)$. As a result, $G \simeq T$ in $K_{\partial U}(\overline{U},C)$ and therefore G is inessential in $K_{\partial U}(\overline{U},C)$ by Theorem 6.2. Symmetry will now imply that F is inessential in $K_{\partial U}(\overline{U},C)$ if and only if G is inessential in $K_{\partial U}(\overline{U},C)$. □

Theorem 6.4 *Let $E = (E, \|\cdot\|)$ be a Banach space, C a closed convex subset of E, U an open subset of C and suppose that $F \in K_{\partial U}(\overline{U},C)$ is essential in $K_{\partial U}(\overline{U},C)$. Then there exists $\epsilon > 0$ with the following properties:*

(i) *any continuous, compact map $G : \overline{U} \to C$ satisfying $\|F(x) - G(x)\| < \epsilon$ for all $x \in \partial U$ belongs to $K_{\partial U}(\overline{U},C)$,*
(ii) *G (as described in (i)) is essential in $K_{\partial U}(\overline{U},C)$.*

Proof Since $F : \overline{U} \to C$ is a compact map and fixed point free on ∂U, an easy argument shows that there exists $\epsilon > 0$ with

$$\|x - F(x)\| \geq \epsilon \text{ for all } x \in \partial U.$$

If $G : \overline{U} \to C$ satisfies

$$\|G(x) - F(x)\| < \epsilon \text{ for all } x \in \partial U,$$

then

$$\|G(x) - x\| \geq \|x - F(x)\| - \|G(x) - F(x)\| > \epsilon - \epsilon = 0$$

and therefore G is fixed point free on ∂U. Apply Theorem 6.1 (or Exercise 6.2) to deduce that $F \simeq G$ in $K_{\partial U}(\overline{U},C)$. Theorem 6.3 now guarantees that G is essential in $K_{\partial U}(\overline{U},C)$. □

We next provide an example of an essential map in $K_{\partial U}(\overline{U},C)$ which is particularly useful in applications.

Theorem 6.5 *Let E be a Banach space, C a closed, convex subset of E, U an open subset of C and $p \in U$. Then the constant map $F(\overline{U}) = p$ is essential in $K_{\partial U}(\overline{U},C)$.*

Proof Let $G : \overline{U} \to C$ be any continuous, compact map with $G|_{\partial U} = F|_{\partial U} = p$. We must show that G has a fixed point in U. Let $J : C \to C$ be given by

$$J(x) := \begin{cases} G(x), & x \in \overline{U}, \\ p, & x \in C \backslash \overline{U}. \end{cases}$$

It is easy to see that $J : C \to C$ is a continuous, compact map. Schauder's fixed point theorem guarantees that J has a fixed point, say $x \in C$. In fact we have $x \in U$ since $p \in U$. Hence $x = J(x) = G(x)$ and therefore x is a fixed point of G. As a result, F is essential in $K_{\partial U}(\overline{U}, C)$. □

Our next result provides an alternative proof of Theorem 5.1.

Theorem 6.6 *Let E be a Banach space, C a closed, convex subset of E, U an open subset of C and $p \in U$. Then every continuous, compact map $F : \overline{U} \to C$ has at least one of the following properties:*

(A1) *F has a fixed point in $\overline{U}$, or*
(A2) *there are a $u \in \partial U$ and a $\lambda \in (0,1)$ with $u = \lambda F(u) + (1 - \lambda)p$.*

Proof Suppose (A2) does not occur and $x \neq F(x)$ for $x \in \partial U$ (otherwise we are finished). Let $G : \overline{U} \to C$ be the constant map $u \mapsto p$ and consider the continuous, compact map $H : \overline{U} \times [0,1] \to C$ joining G to F given by

$$H(x,t) := tF(x) + (1-t)p.$$

In addition, for fixed $t \in [0,1]$, $x \neq H_t(x)$ for $x \in \partial U$. Consequently, $F \simeq G$ in $K_{\partial U}(\overline{U}, C)$. Now Theorem 6.3 and Theorem 6.5 imply that F is essential in $K_{\partial U}(\overline{U}, C)$ and thus there exists $x \in U$ with $x = F(x)$, that is, (A1) occurs. □

Definition 6.5 Let X and Y be two subsets of a Banach space E. A continuous map $f : X \to Y$ is called a *compact field* if $x \mapsto x - f(x)$ is a compact map of X into E.

Definition 6.6 Let (X, d_X) and (Y, d_Y) be metric spaces and $f : X \to Y$. If there are an $\epsilon > 0$ and a $\delta \geq 0$ such that

$$\operatorname{diam} f^{-1}(B(y,\delta)) < \epsilon \text{ for all } y \in Y,$$

then we say that f is a *δ-based ϵ-map;* here $B(y, \delta) := \{z \in Y :$

$d_Y(y,z) < \delta\}$. If $\delta = 0$ the map f is called an *ϵ-map*, and if $\delta > 0$ the map f is called an *ϵ-map in the narrow sense.*

We now use Theorem 6.3 to study the equation $y = x - F(x)$.

Theorem 6.7 *Let $\overline{B}(x_0, \epsilon)$ be the closed ball in the Banach space $E = (E, \|\cdot\|)$ and $f : \overline{B}(x_0, \epsilon) \to E$ be a continuous, compact field. If f is an ϵ-map then there exists an $\eta > 0$ with*

$$B(f(x_0), \eta) \subseteq f(B(x_0, \epsilon)).$$

If f is a δ-based ϵ-map with $\delta > 0$ then

$$B(f(x_0), \delta) \subseteq f(B(x_0, \epsilon)).$$

Proof Let F be the associated compact map, that is, $F(x) := x - f(x)$. Since f is a δ-based ϵ-map, F has the property that

$$\|x - y\| < \epsilon \text{ whenever } \|F(x) - F(y) - (x - y)\| < \delta.$$

Without loss of generality assume that $x_0 = 0$ and therefore $F(0) = -f(0)$. Consider the continuous, compact map G, defined by

$$G(x) := F(x) - F(0),$$

on the ball $\overline{B} = \overline{B}(0, \epsilon)$. We first show that G is essential in $K_{\partial B}(\overline{B}, E)$. To see this, define a map $H : \overline{B} \times [0,1] \to E$ by

$$H(x,t) := F\left(\frac{x}{1+t}\right) - F\left(\frac{-tx}{1+t}\right).$$

It is easy to see that $H : \overline{B} \times [0,1] \to E$ is a continuous, compact map. If

$$F\left(\frac{x}{1+t}\right) - F\left(\frac{-tx}{1+t}\right) = x = \frac{x}{1+t} - \left(\frac{-tx}{1+t}\right)$$

for some $x \in \partial B$ and $t \in [0,1]$, then $\|x\| < \epsilon$ since f is a δ-based ϵ-map. As a result, H_t is fixed point free on ∂B for each $t \in [0,1]$. In addition, since

$$H(x,1) = F\left(\frac{x}{2}\right) - F\left(\frac{-x}{2}\right),$$

we have that $H(-x,1) = -H(x,1)$. Now Exercise 4.11 immediately guarantees that H_1 is essential in $K_{\partial B}(\overline{B}, E)$. This, Theorem 6.3, and the fact that $H_0 \simeq H_1$ in $K_{\partial B}(\overline{B}, E)$, imply that $H_0 = G$ is essential in $K_{\partial B}(\overline{B}, E)$.

We first discuss the case when f is a 0-based ϵ-map. From Theorem 6.4 there is an $\eta > 0$ such that any continuous, compact map G_1 satisfying

$$\|G_1(x) - G(x)\| < \eta \text{ on } \partial B$$

is in $K_{\partial B}(\overline{B}, E)$ and is also essential in $K_{\partial B}(\overline{B}, E)$. Notice in particular that the map $G_1(x) = G(x) + y$ satisfies the following:

for each $\|y\| < \eta$, the equation $x = G(x) + y$ has a solution in B.

This of course says that

$$x - F(x) = -F(0) + y \text{ has a solution in } B \text{ for each } \|y\| < \eta$$

and therefore since $F(0) = -f(0)$ we have that

$$f(x) = f(0) + y \text{ has a solution in } B \text{ for each } \|y\| < \eta.$$

As a result

$$B(f(0), \eta) \subseteq f(B(0, \epsilon)).$$

Now let f be a δ-based ϵ-map with $\delta > 0$. Note that for each $\|y\| < \delta$, the continuous, compact map $N : \overline{B} \times [0,1] \to E$ given by

$$N(x,t) := F(x) - F(0) + ty$$

is such that N_t is fixed point free in ∂B for each $t \in [0,1]$; since if $F(x) - F(0) + ty = x$ for some $x \in \partial B$ and $t \in [0,1]$, then

$$\|F(x) - F(0) - (x - 0)\| \leq t\|y\| < \delta$$

and therefore $\|x\| < \epsilon$. This, Theorem 6.3, and the fact that $G = N_0 \simeq N_1$ in $K_{\partial B}(\overline{B}, E)$ for all $\|y\| < \delta$, imply that N_1 is essential in $K_{\partial B}(\overline{B}, E)$ for all $\|y\| < \delta$. As a result

$$B(f(0), \delta) \subseteq f(B(0, \epsilon)). \qquad \square$$

Theorem 6.8 *Let E be a Banach space and $f : E \to E$ a continuous, completely continuous field. Then*

(a) *if f is an ϵ-map, $f(E)$ is open in E,*
(b) *if f is an ϵ-map in the narrow sense, f is surjective (that is, $f(E) = E$).*

Remark 6.2 Let X and Y be subsets of a Banach space E. A continuous map $f : X \to Y$ is called a completely continuous field if $x \mapsto x - f(x) \equiv F(x)$ is a completely continuous map of X into E (that is, $F(W)$ is relatively compact for all bounded sets $W \subseteq X$).

Proof of Theorem 6.8

(a) We apply Theorem 6.7. Let $V \subseteq E$ be open. We must show that $f(V)$ is open. Take any $z \in f(V)$. Then there exists $x_0 \in V$ with $f(x_0) = z$. Since V is open, there exists $\epsilon > 0$ with $B(x_0, \epsilon) \subseteq V$. Theorem 6.7 guarantees that there is an $\eta > 0$ with

$$B(z, \eta) = B(f(x_0), \eta) \subseteq f(B(x_0, \epsilon)) \subseteq f(V).$$

(b) Theorem 6.7 implies that the ball $B(y, \delta)$ of fixed radius $\delta > 0$ is contained in $f(E)$ for each $y \in f(E)$. As a result, f is surjective. □

Our next result is known as Schauder's domain invariance theorem.

Theorem 6.9 *Let U be an open subset of a Banach space E and $f : U \to E$ an injective, continuous, completely continuous field. Then*

(a) *f is an open map,*
(b) *$f(U)$ is open in E,*
(c) *f is a homeomorphism of U onto $f(U)$.*

Proof Of course since f is injective, it is an ϵ-map for each $\epsilon > 0$. The result now follows from Theorem 6.8 (a). □

An immediate consequence of Theorem 6.9 is the Fredholm alternative.

Theorem 6.10 *Let E be a Banach space and $F : E \to E$ a completely continuous linear operator. Then either*

(a) *the equation $0 = x - F(x)$ has a nontrivial solution, or*
(b) *the equation $y = x - F(x)$ has a unique solution for each $y \in E$.*

Proof The continuous, completely continuous field $f(x) := x - F(x)$ is either injective or not. If f is not injective, then an easy argument shows that (a) holds. On the other hand, if f is injective, then the image of E is a closed, linear subspace. However, Theorem 6.9 implies that $f(E)$ is open in E, therefore $f(E) = E$. Thus the injective field is also surjective and therefore bijective. □

We now discuss condensing maps (see Definition 5.1). In what follows, E will be a Banach space, C a closed, convex subset of E and U an open subset of C.

Definition 6.7 We let $CK(\overline{U}, C)$ denote the set of all continuous, condensing maps $F : \overline{U} \to C$ with $F(\overline{U})$ a bounded set in C.

Definition 6.8 We let $CK_{\partial U}(\overline{U}, C)$ denote the set of maps $F \in CK(\overline{U}, C)$ with $x \neq F(x)$ for $x \in \partial U$.

Definition 6.9 A map $F \in CK_{\partial U}(\overline{U}, C)$ is *essential* in $CK_{\partial U}(\overline{U}, C)$ if for every $G \in CK_{\partial U}(\overline{U}, C)$ with $G|_{\partial U} = F|_{\partial U}$, there exists $x \in U$ with $x = G(x)$. Otherwise F is *inessential* in $CK_{\partial U}(\overline{U}, C)$.

Definition 6.10 Two maps F and G belonging to $CK_{\partial U}(\overline{U}, C)$ are *homotopic* in $CK_{\partial U}(\overline{U}, C)$, written

$$F \simeq G \text{ in } CK_{\partial U}(\overline{U}, C),$$

if there exists a continuous, condensing map $H : \overline{U} \times [0,1] \to C$ with $H(\overline{U} \times [0,1])$ a bounded set in C, together with $H_t(\cdot) : \overline{U} \to C$ belonging to $CK_{\partial U}(\overline{U}, C)$ for each $t \in [0,1]$, $H_0 = F$ and $H_1 = G$; in our definition, $H : \overline{U} \times [0,1] \to C$ *condensing* means $\alpha(H(W)) < \alpha(\Pi W)$ for any bounded subset W of $\overline{U} \times [0,1]$, with $\alpha(W) \neq 0$ (note that $\Pi : \overline{U} \times [0,1] \to \overline{U}$ is the natural projection, that is, the projection onto the $\overline{U}$ coordinate).

We begin with an example of an essential map in $CK_{\partial U}(\overline{U}, C)$.

Theorem 6.11 *Let E be a Banach space, C a closed, convex subset of E, U an open subset of C and $p \in U$. Then the constant map $F(\overline{U}) = p$ is essential in $CK_{\partial U}(\overline{U}, C)$.*

Proof Let $G \in CK_{\partial U}(\overline{U}, C)$ with $G|_{\partial U} = F|_{\partial U} = p$. Define a map $J : C \to C$ by

$$J(x) := \begin{cases} G(x), & x \in \overline{U}, \\ p, & x \in C \backslash \overline{U}. \end{cases}$$

Now $J : C \to C$ is a continuous, condensing map with $J(C)$ a bounded subset of C. The continuity and boundedness part are clear. To show that J is a condensing map, let Ω be a bounded subset of C with $\alpha(\Omega) \neq 0$. Then since

$$J(\Omega) \subseteq G(\Omega \cap \overline{U}) \cup \{p\},$$

we have that

$$\alpha(J(\Omega)) \leq \alpha(G(\Omega \cap \overline{U}) \cup \{p\}) \leq \alpha(G(\Omega)) < \alpha(\Omega),$$

and therefore $J : C \to C$ is a condensing map. We may apply Theorem 4.16 (or Exercise 4.5) to deduce that there exists $x \in C$ with $x = J(x)$. In fact, $x \in U$ since $p \in U$ and therefore $x = J(x) = G(x)$. □

Our next result characterises inessential maps in $CK_{\partial U}(\overline{U}, C)$ in terms of homotopy.

Theorem 6.12 *Let E be a Banach space, C a closed, convex subset of E, U an open subset of C and $F \in CK_{\partial U}(\overline{U}, C)$. Then the following conditions are equivalent:*

(i) *F is inessential in $CK_{\partial U}(\overline{U}, C)$,*
(ii) *there exists a fixed point free map $G \in CK_{\partial U}(\overline{U}, C)$ with $F \simeq G$ in $CK_{\partial U}(\overline{U}, C)$.*

Proof We first show that (i) implies (ii). Let $G \in CK_{\partial U}(\overline{U}, C)$ be a fixed point free map with $F|_{\partial U} = G|_{\partial U}$. Define $H : \overline{U} \times [0,1] \to C$ by

$$H(x,t) := tG(x) + (1-t)F(x).$$

Notice that $H : \overline{U} \times [0,1] \to C$ is a continuous, condensing map with $H(\overline{U} \times [0,1])$ a bounded set in C. The continuity and boundedness of H are clear. To show that H is condensing, let W be a bounded subset of $\overline{U} \times [0,1]$ with $\alpha(W) \neq 0$. Let $\Pi : \overline{U} \times [0,1] \to \overline{U}$ be the natural projection and note that $\alpha(\Pi W) \neq 0$. Notice also that

$$H(W) \subseteq \text{co}(G(\Pi W) \cup F(\Pi W)),$$

and therefore we have

$$\begin{aligned} \alpha(H(W)) &\leq \alpha(\overline{\text{co}}(G(\Pi W) \cup F(\Pi W))) \\ &= \alpha(G(\Pi W) \cup F(\Pi W)) \\ &= \max\{\alpha(G(\Pi W)), \alpha(F(\Pi W))\} < \alpha(\Pi W). \end{aligned}$$

Thus $H : \overline{U} \times [0,1] \to C$ is a condensing map. Also since $F|_{\partial U} = G|_{\partial U}$, we have that H_t is fixed point free on ∂U for each $t \in [0,1]$. In addition, for each $t \in [0,1]$ we have that $H_t : \overline{U} \to C$ is a continuous, condensing map with $H_t(\overline{U})$ a bounded set in C. We need only check that $H_t : \overline{U} \to C$ is a condensing map for each $t \in [0,1]$. Fix $t \in [0,1]$ and let Ω be a bounded subset of $\overline{U}$ with $\alpha(\Omega) \neq 0$. Then $\alpha(\Omega \times \{t\}) \neq 0$ and

$$\alpha(H_t(\Omega)) = \alpha(H(\Omega \times \{t\})) < \alpha(\Pi(\Omega \times \{t\})) = \alpha(\Omega),$$

since $\Pi(\Omega \times \{t\}) = \Omega$. As a result, $H_t \in CK_{\partial U}(\overline{U}, C)$ for each $t \in [0,1]$. Finally, $H_0 = F$ and $H_1 = G$, therefore $F \simeq G$ in $CK_{\partial U}(\overline{U}, C)$.

We now show that (ii) implies (i). Let $H : \overline{U} \times [0,1] \to C$ be a continuous, condensing map with $H(\overline{U} \times [0,1])$ a bounded set in C, $H_t \in CK_{\partial U}(\overline{U}, C)$ for each $t \in [0,1]$, $H_0 = G$ and $H_1 = F$. Consider

$$B := \{x \in \overline{U} : x = H(x,t) \text{ for some } t \in [0,1]\}.$$

If $B = \varnothing$ then in particular $F = H_1$ has no fixed point in $\overline{U}$, therefore F is inessential in $CK_{\partial U}(\overline{U}, C)$. It therefore remains to consider the case when $B \neq \varnothing$. Since B is closed in $\overline{U}$ and $B \cap \partial U = \varnothing$, there exists a continuous function $\mu : \overline{U} \to [0,1]$ with $\mu(\partial U) = 1$ and $\mu(B) = 0$. Define a map $J : \overline{U} \to C$ by

$$J(x) := H(x, \mu(x)).$$

Notice that $J : \overline{U} \to C$ is a continuous, condensing map with $J(\overline{U})$ a bounded set in C. To show that $J : \overline{U} \to C$ is condensing, let Ω be a bounded subset of $\overline{U}$ with $\alpha(\Omega) \neq 0$. Let

$$\Omega^\star := \{(x, \mu(x)) : x \in \Omega\} \subseteq \overline{U} \times [0,1].$$

Note that $\alpha(\Omega^\star) \neq 0$. Then since $J(\Omega) = H(\Omega^\star)$ we have that

$$\alpha(J(\Omega)) = \alpha(H(\Omega^\star)) < \alpha(\Pi(\Omega^\star)) = \alpha(\Omega).$$

In addition, $J|_{\partial U} = F|_{\partial U}$ since if $x \in \partial U$ then

$$J(x) = H(x,1) = F(x).$$

Finally, $x \neq J(x)$ for $x \in \overline{U}$ since $x = J(x)$ for some $x \in \overline{U}$ means that $x \in B$ and therefore $\mu(x) = 0$ (that is, $x = G(x)$) – a contradiction. Consequently, $J \in CK_{\partial U}(\overline{U}, C)$ is fixed point free on $\overline{U}$ with $J|_{\partial U} = F|_{\partial U}$. As a result, F is inessential in $CK_{\partial U}(\overline{U}, C)$. □

Theorem 6.13 *Let E be a Banach space, C a closed, convex subset of E and U an open subset of C. Suppose that F and G are two maps in $CK_{\partial U}(\overline{U}, C)$ with $F \simeq G$ in $CK_{\partial U}(\overline{U}, C)$. Then the map F is essential in $CK_{\partial U}(\overline{U}, C)$ if and only if G is essential in $CK_{\partial U}(\overline{U}, C)$.*

Notes The results in Chapter 6 may be found in Dugundji and Granas [55], Granas [84], Krawcewicz [114] and Precup [154].

Exercises

6.1 Let $E = (E, \|\cdot\|)$ be a Banach space and

$$\overline{B}_r := \{x \in E : \|x - x_0\| \leq r\}.$$

Suppose that $F \in K_{\partial B_r}(\overline{B}_r, E)$ is an essential map with x_0 the only fixed point of F. Show that for every $r_0 \in (0, r)$, the restriction $F|_{\overline{B}_{r_0}}$ is essential in $K_{\partial B_{r_0}}(\overline{B}_{r_0}, E)$.

6.2 Let $E = (E, \|\cdot\|)$ be a Banach space, C a closed, convex subset of E, U an open subset of C and $F, G \in K_{\partial U}(\overline{U}, C)$ with

$$\sup_{x \in \partial U} \|F(x) - G(x)\| \leq \inf_{x \in \partial U} \|x - F(x)\|.$$

Show that $F \simeq G$ in $K_{\partial U}(\overline{U}, C)$.

6.3 Let $f : E \to E$ be a continuous, completely continuous field in a Banach space $E = (E, \|\cdot\|)$. If

$$\|f(x) - f(y)\| \geq M\|x - y\| \text{ for some } M > 0,$$

show that f is a homeomorphism of E onto E.

6.4 Let E be a Banach space and C a closed subset of E. Show that the continuous, compact field $f : C \to E$ is a closed, proper map, that is, show that the image of each closed set in C is closed in E, and the preimage of each compact subset of E is compact.

6.5 Let U be a bounded, open neighbourhood of 0 in an infinite dimensional Banach space $E = (E, \|\cdot\|)$ and let $F : \partial U \to E$ be a continuous, compact map. Suppose that there exists $\alpha > 0$ with $\|F(x)\| \geq \alpha$ for all $x \in \partial U$. Show that there exist $x \in \partial U$ and $\mu > 0$ with $x = \mu F(x)$.

6.6 Let E be a Banach space, C a closed, convex subset of E, U an open, bounded subset of C with $0 \in U$ and $F : \overline{U} \to C$ a continuous, compact map.

(a) Suppose that $F(x) \neq \lambda x$ for all $x \in \partial U$ and $\lambda \geq 1$. Show that F is essential in $K_{\partial U}(\overline{U}, C)$.

(b) Suppose that there exists $c \in C$, with $c \neq 0$ and $x - F(x) \neq \lambda c$ for all $x \in \partial U$ and $\lambda \geq 0$. Show that the map F is inessential in $K_{\partial U}(\overline{U}, C)$.

6.7 Show that it is enough to assume that E is a normed linear space in Theorem 6.3, Theorem 6.5 and Theorem 6.6.

6.8 Let U be an open set in a Banach space E and $f : U \to E$ be a map of the form

$$f(x) = x - [F(x) + G(x)]$$

where G is contractive and F is continuous and completely continuous. If f is injective show that

(i) f is open,

(ii) $f : U \to f(U)$ is a homeomorphism.

7
Fixed Point Theorems in Conical Shells

In this chapter we present Krasnoselskii's compression and expansion of a cone theorem and some generalisations. These results can be particularly useful in establishing the existence of multiple solutions of operator equations.

Throughout the chapter $E = (E, \|\cdot\|)$ will denote a Banach space and C will be a closed, convex, nonempty subset of E with $\alpha u + \beta v \in C$ for all $\alpha \geq 0$, $\beta \geq 0$ and $u, v \in C$. Let $\rho > 0$ with

$$B_\rho := \{x : x \in C \text{ and } \|x\| < \rho\}, \quad S_\rho := \{x : x \in C \text{ and } \|x\| = \rho\}$$

and of course $\overline{B}_\rho := B_\rho \cup S_\rho$.

Theorem 7.1 *Let E and C be as described above and let r, R be constants with $0 < r < R$. Suppose that $F \in K(\overline{B}_R, C)$ and assume the following conditions hold:*

$$x \neq F(x) \text{ for } x \in S_r \cup S_R, \tag{7.1}$$

$$\begin{cases} F : \overline{B}_r \to C \text{ is inessential in } K_{S_r}(\overline{B}_r, C), \\ \text{that is, } F|_{\overline{B}_r} \text{ is inessential in } K_{S_r}(\overline{B}_r, C), \end{cases} \tag{7.2}$$

and

$$F : \overline{B}_R \to C \text{ is essential in } K_{S_R}(\overline{B}_R, C). \tag{7.3}$$

Then F has at least one fixed point in $\Omega := \{x : x \in C$ and $r < \|x\| < R\}$.

Proof Suppose that F has no fixed points in Ω. Condition (7.2) implies that there exists $\theta \in K(\overline{B}_r, C)$ with $\theta|_{S_r} = F|_{S_r}$ and $x \neq \theta(x)$ for

$x \in \overline{B}_r$. Define the mapping $\Phi : \overline{B}_R \to C$ by

$$\Phi(x) := \begin{cases} F(x), & r < \|x\| \le R, \\ \theta(x), & 0 \le \|x\| \le r. \end{cases}$$

Notice that $\Phi \in K(\overline{B}_R, C)$ and in addition, Φ has no fixed points in $\overline{B}_R$ (since θ has no fixed points in $\overline{B}_r$ and F has no fixed points in Ω). This contradicts the fact that $F : \overline{B}_R \to C$ is essential in $K(\overline{B}_R, C)$. □

Theorem 7.2 *Let E and C be as described above and let r, R be constants with $0 < r < R$. Suppose that the following conditions are satisfied:*

$$(7.4)\qquad \begin{cases} N : \overline{B}_R \times [0,1] \to C \text{ is a continuous, compact map,} \\ \text{with } N(x,0) = 0 \text{ for all } x \in \overline{B}_R, \text{and such that for each} \\ t \in [0,1], \text{ we have that } x \ne N(x,t) \text{ for all } x \in S_R, \end{cases}$$

$$(7.5)\qquad \begin{cases} H : \overline{B}_r \times [0,1] \to C \text{ is a continuous, compact map, such} \\ \text{that for each } t \in [0,1], \text{ we have } x \ne H(x,t) \text{ for all } x \in S_r, \end{cases}$$

$$(7.6)\qquad H(\cdot,1)|_{\overline{B}_r} = N(\cdot,1)|_{\overline{B}_r}$$

and

$$(7.7)\qquad x \ne H(x,0) \text{ for all } x \in B_r.$$

Then $N(\cdot,1)$ has a fixed point in $\Omega := \{x : x \in C$ and $r < \|x\| < R\}$.

Proof Note that Theorem 6.5 guarantees that the zero map is essential in $K_{S_R}(\overline{B}_R, C)$. Condition (7.4) together with Theorem 6.3 implies that

$$(7.8)\qquad N(\cdot,1) : \overline{B}_R \to C \text{ is essential in } K_{S_R}(\overline{B}_R, C).$$

In addition, (7.7) (and also (7.5)) implies that $H(\cdot,0)$ is inessential in $K_{S_r}(\overline{B}_r, C)$. This together with (7.5), (7.6) and Theorem 6.3 gives

$$(7.9)\qquad N(\cdot,1) = H(\cdot,1) : \overline{B}_r \to C \text{ is inessential in } K_{S_r}(\overline{B}_r, C).$$

Now (7.8), (7.9) (and also (7.4), (7.5) and (7.6)) imply that (7.1), (7.2) and (7.3) of Theorem 7.1 hold with $F(\cdot) = N(\cdot,1)$. The result now follows from Theorem 7.1. □

We now prove Krasnoselskii's compression of a cone theorem.

Theorem 7.3 *Let E and C be as described above and let r, R be constants with $0 < r < R$. Suppose that $F \in K(\overline{B}_R, C)$ and assume the following conditions hold:*

$$x \neq \lambda F(x) \text{ for } \lambda \in [0,1) \text{ and } x \in S_R \tag{7.10}$$

and

$$\begin{cases} \text{there exists a } v \in C\backslash\{0\} \text{ with} \\ x \neq F(x) + \delta v \text{ for any } \delta > 0 \text{ and } x \in S_r. \end{cases} \tag{7.11}$$

Then F has a fixed point in $\overline{\Omega} = \{x : x \in C$ and $r \leq \|x\| \leq R\}$.

Proof Suppose that $x \neq F(x)$ for $x \in S_r \cup S_R$ (otherwise we are finished). Choose $M > 0$ such that

$$\|F(x)\| \leq M \text{ for all } x \in \overline{B}_r.$$

Next choose $\delta_0 > 0$ such that

$$\|\delta_0 v\| > M + r. \tag{7.12}$$

Let

$$N(\cdot, t) := tF(\cdot) \text{ and } H(\cdot, t) := F(\cdot) + (1-t)\delta_0 v.$$

Conditions (7.10) and (7.11) (with $\delta = (1-t)\delta_0$) imply that (7.4) and (7.5) are satisfied. In addition, (7.6) is true since

$$N(x, 1) = F(x) = H(x, 1) \text{ for } x \in \overline{B}_r.$$

Finally, (7.12) implies that (7.7) is satisfied (note $H(x, 0) = F(x) + \delta_0 v$). The result now follows from Theorem 7.2. □

In the next theorem $C \subseteq E$ will be a cone. Let $\rho > 0$ with

$$\Omega_\rho := \{x \in E : \|x\| < \rho\}, \quad \partial_E \Omega_\rho := \{x \in E : \|x\| = \rho\},$$

$$B_\rho := \{x : x \in C \text{ and } \|x\| < \rho\} \text{ and } S_\rho := \{x : x \in C \text{ and } \|x\| = \rho\}.$$

Notice that

$$B_\rho = \Omega_\rho \cap C \text{ and } S_\rho = \partial_E(\Omega_\rho \cap C) = \partial_E \Omega_\rho \cap C.$$

Theorem 7.4 *Let $E = (E, \|\cdot\|)$ be a Banach space, $C \subseteq E$ a cone and let $\|\cdot\|$ be increasing with respect to C. In addition, let r, R be constants with $0 < r < R$. Suppose that $F : \overline{\Omega}_R \cap C \to C$ is a continuous, compact map and assume the following conditions hold:*

$$\|F(x)\| \leq \|x\| \text{ for all } x \in \partial_E \Omega_R \cap C \tag{7.13}$$

and

(7.14) $$\|F(x)\| > \|x\| \text{ for all } x \in \partial_E \Omega_r \cap C.$$

Then F has a fixed point in $C \cap \{x \in E : r \leq \|x\| \leq R\}$.

Proof Notice that (7.13) implies that (7.10) is true. To see this, suppose there exist $x \in S_R$ and $\lambda \in [0, 1)$ with $x = \lambda F(x)$. Then

$$R = \|x\| = |\lambda| \, \|F(x)\| < \|F(x)\| \leq \|x\| = R$$

– a contradiction. Also (7.14) implies that (7.11) is true. To see this, suppose there exists $v \in C \backslash \{0\}$ with $x = F(x) + \delta v$ for some $\delta > 0$ and $x \in S_r$. Since $\| \cdot \|$ is increasing with respect to C we have, since $\delta v \in C$,

$$\|x\| = \|F(x) + \delta v\| \geq \|F(x)\| > \|x\|$$

– a contradiction. The result now follows from Theorem 7.3. □

For our next two results we will again assume that $C \subseteq E$ is a closed, convex, nonempty set with $\alpha u + \beta v \in C$ for all $\alpha \geq 0$, $\beta \geq 0$ and u, $v \in C$.

Theorem 7.5 *Let E be a Banach space and $C \subseteq E$ a closed, convex, nonempty set with $\alpha u + \beta v \in C$ for all $\alpha \geq 0$, $\beta \geq 0$ and u, $v \in C$. In addition, let r, R be constants with $0 < r < R$. Suppose that $F \in K(\overline{B}_R, C)$ and assume the following conditions hold:*

(7.15) $$x \neq F(x) \text{ for } x \in S_r \cup S_R,$$

(7.16) $$\begin{cases} F : \overline{B}_r \to C \text{ is essential in } K_{S_r}(\overline{B}_r, C), \\ \text{that is, } F|_{\overline{B}_r} \text{ is essential in } K_{S_r}(\overline{B}_r, C), \end{cases}$$

and

(7.17) $$F : \overline{B}_R \to C \text{ is inessential in } K_{S_R}(\overline{B}_R, C).$$

Then F has at least two fixed points x_0 and x_1 with $x_0 \in B_r$ and $x_1 \in \Omega = \{x : x \in C$ and $r < \|x\| < R\}$.

Proof We know from (7.16) that F has a fixed point in B_r.

Let $\Psi := F|_{\overline{\Omega}}$, and suppose that $\Psi : \overline{\Omega} \to C$ has no fixed points. The fact that $F : \overline{B}_R \to C$ is inessential in $K_{S_R}(\overline{B}_R, C)$ means that there exists a continuous, compact map $\theta : \overline{B}_R \to C$ with $\theta|_{S_R} = F|_{S_R}$ and $x \neq \theta(x)$ for $x \in \overline{B}_R$. Fix $\rho \in (0, r)$ and consider the map Φ given by

$$\Phi(x) :=$$

$$\begin{cases} \dfrac{\rho}{R}\theta\left(\dfrac{R}{\rho}x\right), & 0 \le \|x\| \le \rho, \\[2ex] \dfrac{(r-\rho)\|x\|}{(R-\rho)r-(R-r)\|x\|}\Psi\left(\dfrac{(R-\rho)r-(R-r)\|x\|}{(r-\rho)\|x\|}x\right), & \rho \le \|x\| \le r, \\[2ex] \Psi(x), & r \le \|x\| \le R. \end{cases}$$

Notice that $\Phi : \overline{B}_R \to C$ is well defined since if $\rho \le \|x\| \le r$, then

$$r \le \left\| \frac{(R-\rho)r-(R-r)\|x\|}{(r-\rho)\|x\|}x \right\| \le R.$$

Also $\Phi : \overline{B}_R \to C$ is a continuous, compact map. In addition,

$$\Phi|_{S_R} = \Psi|_{S_R} = F|_{S_R} \text{ and } \Phi|_{\overline{\Omega}} = \Psi|_{\overline{\Omega}} = F|_{\overline{\Omega}},$$

and therefore Φ has no fixed points in $\overline{B}_R$ (since θ has no fixed points in $\overline{B}_R$ and F has no fixed points in $\overline{\Omega}$).

Let us concentrate on $\Phi : \overline{B}_r \to C$ (that is, $\Phi|_{\overline{B}_r}$). Now

$$\Phi|_{S_r} = \Psi|_{S_r} = F|_{S_r},$$

therefore $\Phi : \overline{B}_r \to C$ is a continuous, compact map with $\Phi|_{S_r} = F|_{S_r}$ and Φ has no fixed points in $\overline{B}_r$. This contradicts (7.16). □

Next we prove Krasnoselskii's expansion of a cone theorem.

Theorem 7.6 *Let E be a Banach space and $C \subseteq E$ a closed, convex, nonempty set with $\alpha u + \beta v \in C$ for all $\alpha \ge 0$, $\beta \ge 0$ and u, $v \in C$. In addition, let r, R be constants with $0 < r < R$. Suppose that $F \in K(\overline{B}_R, C)$ and assume the following conditions hold:*

$$x \ne \lambda F(x) \text{ for } \lambda \in [0,1) \text{ and } x \in S_r \tag{7.18}$$

and

$$\begin{cases} \text{there exists a } v \in C\backslash\{0\} \text{ with} \\ x \ne F(x) + \delta v \text{ for any } \delta > 0 \text{ and } x \in S_R. \end{cases} \tag{7.19}$$

Then F has a fixed point in $\overline{\Omega} = \{x : x \in C$ and $r \le \|x\| \le R\}$.

Proof Assume that $x \ne F(x)$ for $x \in S_r \cup S_R$ (otherwise we are finished). The result follows immediately from Theorem 7.5 once we show that conditions (7.16) and (7.17) are satisfied.

Consider the homotopy $H : \overline{B}_r \times [0, 1] \to C$ defined by

$$H(x, \lambda) := \lambda F(x).$$

Notice that $H_0 = 0$, $H_1 = F$ and therefore since $x \neq F(x)$ on S_r, we have $H_0 \simeq H_1$ in $K_{S_r}(\overline{B}_r, C)$ (this follows since $H : \overline{B}_r \times [0, 1] \to C$ is a continuous, compact map with $H_t \in K_{S_r}(\overline{B}_r, C)$ for each $t \in [0, 1]$ since (7.18) holds). From Theorem 6.5 we have that $H_0 : \overline{B}_r \to C$ is essential in $K_{S_r}(\overline{B}_r, C)$, and this together with Theorem 6.3 implies that $F : \overline{B}_r \to C$ is essential in $K_{S_r}(\overline{B}_r, C)$. Thus (7.16) holds.

Let $\delta_0 > 0$ be such that

$$\|\delta_0 v\| > \sup_{x \in S_R} \|F(x)\| + R. \tag{7.20}$$

Consider the homotopy $N : \overline{B}_R \times [0, 1] \to C$ defined by

$$N(x, \lambda) := F(x) + \lambda \delta_0 v.$$

Notice that N is a continuous, compact map with $N_0 = F$ and $N_1 = F + \delta_0 v$. Then since (7.19) holds (for all $\delta \geq 0$, since we are assuming that $x \neq F(x)$ on S_R), we have

$$N_0 \simeq N_1 \text{ in } K_{S_R}(\overline{B}_R, C). \tag{7.21}$$

Notice (7.20) implies for $\lambda \in [0, 1]$ and $x \in S_R$ that

$$\|\delta_0 v + \lambda F(x)\| > R = \|x\|.$$

Thus

$$x \neq \lambda F(x) + \delta_0 v \text{ if } \lambda \in [0, 1] \text{ and } x \in S_R. \tag{7.22}$$

Let $G : \overline{B}_R \to C$ be the constant map $G(x) := \delta_0 v$. Consider the homotopy $J : \overline{B}_R \times [0, 1] \to C$ defined by

$$J(x, \lambda) := \delta_0 v + \lambda F(x).$$

Notice J is a continuous, compact map with $J_0 = G$ and $J_1 = N_1$. From (7.22) we have that

$$N_1 \simeq G \text{ in } K_{S_R}(\overline{B}_R, C). \tag{7.23}$$

Now (7.21) and (7.23) imply

$$N_0 \simeq G \text{ in } K_{S_R}(\overline{B}_R, C). \tag{7.24}$$

However since $G(x) = \delta_0 v$ for $x \in \overline{B}_R$ and $\delta_0 v \in C \backslash \overline{B}_R$ (see (7.20)) it is immediate that G is inessential in $K_{S_R}(\overline{B}_R, C)$. Theorem 6.3 now implies that $N_0 = F$ is inessential in $K_{S_R}(\overline{B}_R, C)$. Thus (7.17) holds. □

Theorem 7.7 *Let $E = (E, \|\cdot\|)$ be a Banach space, $C \subseteq E$ a cone and let $\|\cdot\|$ be increasing with respect to C. In addition, let r, R be constants with $0 < r < R$. Suppose that $F : \overline{\Omega}_R \cap C \to C$ (here $\Omega_R := \{x \in E : \|x\| < R\}$) is a continuous, compact map and assume the following conditions hold:*

(7.25) $$\|F(x)\| > \|x\| \text{ for all } x \in \partial_E \Omega_R \cap C$$

and

(7.26) $$\|F(x)\| \leq \|x\| \text{ for all } x \in \partial_E \Omega_r \cap C.$$

Then F has a fixed point in $C \cap \{x \in E : r \leq \|x\| \leq R\}$.

Proof Notice that (7.25) and (7.26) imply that (7.18) and (7.19) are true (see the ideas used in Theorem 7.4). □

We now combine some of the theorems in this chapter (and also Chapter 6) to establish the existence of multiple fixed points.

Theorem 7.8 *Let $E = (E, \|\cdot\|)$ be a Banach space, $C \subseteq E$ a cone and let $\|\cdot\|$ be increasing with respect to C. In addition, let r, R be constants with $0 < r < R$. Suppose that $F : \overline{\Omega}_R \cap C \to C$ (here $\Omega_R := \{x \in E : \|x\| < R\}$) is a continuous, compact map and assume the following conditions hold:*

(7.27) $$x \neq F(x) \text{ for all } x \in \partial_E \Omega_r \cap C,$$

(7.28) $$\|F(x)\| > \|x\| \text{ for all } x \in \partial_E \Omega_R \cap C$$

and

(7.29) $$\|F(x)\| \leq \|x\| \text{ for all } x \in \partial_E \Omega_r \cap C.$$

Then F has at least two fixed points x_0 and x_1 with $x_0 \in \Omega_r \cap C$ and $x_1 \in C \cap (\overline{\Omega}_R \backslash \overline{\Omega}_r)$.

Proof Theorem 6.6 (note (7.29) implies that $x \neq \lambda F(x)$ for all $\lambda \in [0, 1)$ and $x \in \partial_E \Omega_r \cap C$) implies that F has a fixed point $x_0 \in C \cap \overline{\Omega}_r$. In fact, (7.27) implies that $x_0 \in \Omega_r \cap C$. In addition, Theorem 7.7 implies that F has a fixed point $x_1 \in C \cap (\overline{\Omega}_R \backslash \Omega_r)$. In fact, $x_1 \in C \cap (\overline{\Omega}_R \backslash \overline{\Omega}_r)$.

(To see this, note that if this is not true, then $x_1 \in C$ and $\|x_1\| = r$, therefore $x_1 \in \partial_E\Omega_r \cap C$. This contradicts (7.27).) □

Theorem 7.9 *Let $E = (E, \|\cdot\|)$ be a Banach space, $C \subseteq E$ a cone and let $\|\cdot\|$ be increasing with respect to C. In addition, let L, r, R be constants with $0 < L < r < R$. Suppose that $F : \overline{\Omega}_R \cap C \to C$ is a continuous, compact map and assume the following conditions hold:*

(7.30) $$x \neq F(x) \text{ for all } x \in \partial_E\Omega_r \cap C,$$

(7.31) $$\|F(x)\| > \|x\| \text{ for all } x \in \partial_E\Omega_L \cap C,$$

(7.32) $$\|F(x)\| \leq \|x\| \text{ for all } x \in \partial_E\Omega_r \cap C$$

and

(7.33) $$\|F(x)\| > \|x\| \text{ for all } x \in \partial_E\Omega_R \cap C.$$

Then F has at least two fixed points x_0 and x_1 with $x_0 \in C \cap (\Omega_r \backslash \Omega_L)$ and $x_1 \in C \cap (\overline{\Omega}_R \backslash \overline{\Omega}_r)$.

Proof Theorem 7.4 implies that F has a fixed point $x_0 \in C \cap (\overline{\Omega}_r \backslash \Omega_L)$. In fact (7.30) guarantees that $x_0 \in C \cap (\Omega_r \backslash \Omega_L)$. Theorem 7.7 guarantees the existence of $x_1 \in C \cap (\overline{\Omega}_R \backslash \Omega_r)$. However, (7.30) guarantees that $x_1 \in C \cap (\overline{\Omega}_R \backslash \overline{\Omega}_r)$. □

We now illustrate how the ideas and results in this chapter can be applied in practice. Our goal is to prove the existence of multiple solutions of the nonlinear Fredholm integral equation

(7.34) $$y(t) = h(t) + \int_0^1 k(t,s) f(y(s))\, ds \text{ for } t \in [0,1].$$

In particular we want to establish the existence of *multiple nonnegative* solutions of (7.34). However before tackling this problem, it is reasonable to first consider what conditions one requires on h, k and f in order for (7.34) to have at least *one nonnegative* solution $y \in C[0,1]$.

Theorem 7.10 *Suppose that*

(7.35) $$0 \leq k_t(s) = k(t,s) \in L^1[0,1] \text{ for each } t \in [0,1],$$

(7.36) $$\text{the map } t \mapsto k_t \text{ is continuous from } [0,1] \text{ to } L^1[0,1],$$

(7.37) $\begin{cases} \textit{there exist } 0 < M < 1,\ \kappa \in L^1[0,1] \textit{ and an} \\ \textit{interval } [a,b] \subseteq [0,1],\ a < b, \textit{ such that} \\ k(t,s) \geq M\kappa(s) \geq 0, \quad \textit{for } t \in [a,b],\ \textit{a.e. } s \in [0,1], \end{cases}$

(7.38) $k(t,s) \leq \kappa(s),\ t \in [0,1], \quad \textit{a.e. } s \in [0,1],$

(7.39) $\begin{cases} h \in C[0,1] \textit{ with } h(t) \geq 0 \textit{ for } t \in [0,1], \quad \textit{and} \\ \min_{t\in[a,b]} h(t) \geq M|h|_0 = M \sup_{t\in[0,1]} |h(t)|, \end{cases}$

(7.40) $\begin{cases} f : \mathbf{R} \to \mathbf{R} \textit{ is continuous and nondecreasing} \\ \textit{with } f(u) > 0 \textit{ for } u > 0, \end{cases}$

(7.41) $\begin{cases} \textit{there exists } \alpha > 0 \textit{ such that } \dfrac{\alpha}{|h|_0 + K_1 f(\alpha)} > 1 \\ \textit{where } K_1 = \sup_{t\in[0,1]} \displaystyle\int_0^1 k(t,s)\,ds > 0 \end{cases}$

and

(7.42) $\begin{cases} \textit{there exist } \beta > 0,\ \beta \neq \alpha, \textit{ and } t^\star \in [0,1] \textit{ such that} \\ \dfrac{\beta}{h(t^\star) + f(M\beta)\displaystyle\int_a^b k(t^\star, s)\,ds} < 1 \end{cases}$

hold. Then (7.34) *has at least one nonnegative solution* $y \in C[0,1]$ *and either*

(A) $0 < \alpha < |y|_0 < \beta$ *and* $y(t) \geq M\alpha,\ t \in [a,b],$ *if* $\alpha < \beta$

or

(B) $0 < \beta < |y|_0 < \alpha$ *and* $y(t) \geq M\beta,\ t \in [a,b],$ *if* $\beta < \alpha$

holds.

Proof Define the operator $F : C[0,1] \to C[0,1]$ by

(7.43) $$Fy(t) := h(t) + \int_0^1 k(t,s) f(y(s))\,ds, \quad t \in [0,1],$$

the cone C by

$$C := \left\{ y \in C[0,1] : y(t) \geq 0 \text{ for } t \in [0,1] \text{ and } \min_{t\in[a,b]} y(t) \geq M|y|_0 \right\};$$

and let

$$\Omega_\alpha := \{y \in C[0,1] : |y|_0 < \alpha\}$$

and

$$\Omega_\beta := \{y \in C[0,1] : |y|_0 < \beta\}.$$

Conditions (7.35), (7.36), (7.39) and (7.40) imply that (see Theorem 5.2)

$F : C[0,1] \to C[0,1]$ is continuous and completely continuous.

Also for $y \in C[0,1]$ with $y(t) \geq 0$ for $t \in [0,1]$, we have from (7.35), (7.38), (7.39) and (7.40) that

$$\text{(7.44)} \qquad \begin{aligned} |Fy|_0 &= \sup_{t\in[0,1]} \left(h(t) + \int_0^1 k(t,s) f(y(s))\, ds \right) \\ &\leq |h|_0 + \int_0^1 \kappa(s) f(y(s))\, ds \end{aligned}$$

holds. Therefore for $y \in C[0,1]$ with $y(t) \geq 0$ for $t \in [0,1]$, this fact along with (7.37) and (7.39) yields

$$\begin{aligned} \min_{t\in[a,b]} Fy(t) &= \min_{t\in[a,b]} \left(h(t) + \int_0^1 k(t,s) f(y(s))\, ds \right) \\ &\geq M \left(|h|_0 + \int_0^1 \kappa(s) f(y(s))\, ds \right) \\ &\geq M|Fy|_0. \end{aligned}$$

Consequently we have that

$F : C \to C$ is continuous and completely continuous.

The desired result will now follow from either Theorem 7.4 or Theorem 7.7 if we show that

$$\text{(7.45)} \qquad |Fy|_0 < |y|_0 \text{ for } y \in \partial\Omega_\alpha \cap C$$

and

$$\text{(7.46)} \qquad |Fy|_0 > |y|_0 \text{ for } y \in \partial\Omega_\beta \cap C$$

are true.

We first show that (7.45) holds. Let $y \in \partial\Omega_\alpha \cap C$. In particular this means that $y(t) \geq 0$ for $t \in [0,1]$ and $|y|_0 = \alpha$. Then from (7.35), (7.39), (7.40) and (7.41) we obtain

$$\begin{aligned}
\sup_{t\in[0,1]} |Fy(t)| &= \sup_{t\in[0,1]} \left(h(t) + \int_0^1 k(t,s) f(y(s))\, ds \right) \\
&\leq \sup_{t\in[0,1]} h(t) + f(|y|_0) \sup_{t\in[0,1]} \int_0^1 k(t,s)\, ds \\
&\leq |h|_0 + K_1 f(\alpha) < \alpha = |y|_0.
\end{aligned}$$

Therefore (7.45) is true.

We secondly show that (7.46) is true. Let $y \in \partial\Omega_\beta \cap C$. Then $y(t) \geq 0$ for $t \in [0,1]$, $|y|_0 = \beta$ and for $t \in [a,b]$ we have $M\beta \leq y(t) \leq \beta$. Now (7.35), (7.39), (7.40) and (7.42) imply that

$$\begin{aligned}
Fy(t^\star) &= h(t^\star) + \int_0^1 k(t^\star, s) f(y(s))\, ds \\
&\geq h(t^\star) + \int_a^b k(t^\star, s) f(y(s))\, ds \\
&\geq h(t^\star) + f(M\beta) \int_a^b k(t^\star, s)\, ds \\
&> \beta = |y|_0.
\end{aligned}$$

Consequently (7.46) holds.

Hence by Theorem 7.4 or Theorem 7.7 there exists $y \in C$ that solves (7.34). In particular, if $\alpha < \beta$ then Theorem 7.7 implies that $y \in C \cap (\overline{\Omega}_\beta \backslash \Omega_\alpha)$, while if $\beta < \alpha$ then Theorem 7.4 implies that in fact $y \in C \cap (\overline{\Omega}_\alpha \backslash \Omega_\beta)$. □

Remark 7.1 Condition (7.42) may be replaced by

$$\text{(7.47)} \qquad \begin{cases} \text{there exists } \beta > 0, \ \ \beta \neq \alpha, \ \text{ such that} \\ \dfrac{\beta}{K_2 f(M\beta)} < 1 \text{ where } K_2 = \sup\limits_{t\in[0,1]} \displaystyle\int_a^b k(t,s)\, ds, \end{cases}$$

since (7.46) also follows from (7.47). To see this note that for $y \in \partial\Omega_\beta \cap C$, (7.35), (7.39), (7.40) and (7.47) give

$$\begin{aligned}
\sup_{t\in[0,1]} |Fy(t)| &= \sup_{t\in[0,1]} \left(h(t) + \int_0^1 k(t,s) f(y(s))\, ds \right) \\
&\geq \sup_{t\in[0,1]} \int_0^1 k(t,s) f(y(s))\, ds
\end{aligned}$$

$$\geq \quad K_2 f(M\beta) > \beta = |y|_0.$$

Hence (7.47) implies that (7.46) is true. In particular if $|h|_0 = 0$, it is better to replace (7.42) with (7.47).

Remark 7.2 It is easy to check that the requirement that $f : \mathbf{R} \to \mathbf{R}$ (as described in (7.40)) be nondecreasing can be omitted, provided that (7.41) and (7.42) be replaced with

$$\text{(7.48)} \qquad \text{there exists } \alpha > 0 \text{ such that } \frac{\alpha}{|h|_0 + K_1 \sup\limits_{z\in[0,\alpha]} f(z)} > 1$$

and

$$\text{(7.49)} \qquad \begin{cases} \text{there exist } \beta > 0, \;\; \beta \neq \alpha, \text{ and } t^\star \in [0,1] \text{ such that} \\ \dfrac{\beta}{h(t^\star) + \inf\limits_{z\in[M\beta,\beta]} f(z) \displaystyle\int_a^b k(t^\star, s)\, ds} < 1 \end{cases}$$

respectively.

Example 7.1 Consider the separable kernel

$$k(t,s) := k_1(t)k_2(s), \;\; t \in [0,1], \text{ a.e. } s \in [0,1],$$

where $k_1 \in C[0,1]$ is such that $0 \leq k_1(t) < 1$, $t \in [0,1]$, with $k_1(t) > 0$ on some interval $[a,b] \subseteq [0,1]$, and $k_2 \in L^1[0,1]$ is such that $k_2(s) \geq 0$ for a.e. $s \in [0,1]$. Then k satisfies (7.35)–(7.38) with $\kappa = k_2$ and $M = \min\limits_{t\in[a,b]} k_1(t)$.

Extracting some of the hypotheses of Theorem 7.10 we have the following existence result which also guarantees the existence of at least one nonnegative solution $y \in C[0,1]$ of (7.34). In this result however, it is possible that $y \equiv 0$ may be the solution, whereas in Theorem 7.10 we have that the guaranteed solution is positive on some interval $[a,b] \subseteq [0,1]$.

Theorem 7.11 *Suppose that* (7.35), (7.36), (7.40), (7.41) *and*

$$\text{(7.50)} \qquad h \in C[0,1] \textit{ with } h(t) \geq 0 \textit{ for } t \in [0,1]$$

hold. Then (7.34) *has at least one nonnegative solution* $y \in C[0,1]$ *such that* $0 \leq y(t) < \alpha$ *for* $t \in [0,1]$.

Proof Define

$$f^\star(y) := \begin{cases} f(y), & y \geq 0, \\ f(0), & y < 0, \end{cases}$$

and consider the integral equation

$$y(t) = h(t) + \int_0^1 k(t,s) f^\star(y(s))\, ds \text{ for } t \in [0,1]. \tag{7.51}$$

As in Theorem 5.2, the operator $F^\star : C[0,1] \to C[0,1]$ given by

$$F^\star y(t) := h(t) + \int_0^1 k(t,s) f(y(s))\, ds, \quad t \in [0,1],$$

is continuous and completely continuous. Let $y \in C[0,1]$ be any solution of

$$y(t) = \lambda \left(h(t) + \int_0^1 k(t,s) f^\star(y(s))\, ds \right) \text{ for } t \in [0,1], \text{ some } \lambda \in (0,1).$$

Certainly $y(t) \geq 0$ for $t \in [0,1]$. In addition (7.40) implies that

$$\begin{aligned} |y|_0 &= \sup_{t \in [0,1]} \left(h(t) + \int_0^1 k(t,s) f(y(s))\, ds \right) \\ &\leq |h|_0 + K_1 f(|y|_0), \end{aligned}$$

and consequently we have from (7.41) that $|y|_0 \neq \alpha$.

Applying Theorem 5.1 with $C = E = C[0,1]$,

$$U := \{ y \in C[0,1] : |y|_0 < \alpha \} \text{ and } F = F^\star,$$

we see that (7.51) has at least one solution $y \in C[0,1]$ such that $0 \leq |y|_0 < \alpha$. However since $y(t) \geq 0$ for $t \in [0,1]$, y is also a solution of (7.34) and the result is proved. □

Remark 7.3 If h, k and f are such that the hypotheses of Theorem 7.11 hold and in addition $y \equiv 0$ is *not* a solution of (7.34), then Theorem 7.11 implies that (7.34) has at least one solution $y \in C[0,1]$ that satisfies $0 < |y|_0 < \alpha$.

Returning to Theorem 7.10 we note that conditions (7.41) and (7.42) indicate the central role played by the nonlinearity f in establishing the existence of at least one nonnegative solution $y \in C[0,1]$ of (7.34). By imposing further conditions of this type on f we now obtain the following multiple solutions result for (7.34).

Theorem 7.12 *Suppose that (7.35)–(7.40),*

$$(7.52)\qquad \begin{cases} \textit{there exist constants } \alpha_i > 0,\ i = 1,\dots,n \textit{ for} \\ \textit{some } n \in \mathbf{N} = \{1,2,\dots\},\ \textit{ such that for each} \\ i \in \{1,\dots,n\},\ (7.41) \textit{ is satisfied with } \alpha = \alpha_i \end{cases}$$

and

$$(7.53)\qquad \begin{cases} \textit{there exist constants } \beta_j > 0 \textit{ and } t_j^\star \in [0,1],\ j = 1,\dots,m, \\ \textit{some } m \in \mathbf{N},\ \textit{ such that for each } j \in \{1,\dots,m\}, \\ (7.42) \textit{ is satisfied with } \beta = \beta_j \textit{ and } t^\star = t_j^\star \end{cases}$$

hold.

(I) *If $m = n+1$ and $0 < \beta_1 < \alpha_1 < \cdots < \beta_n < \alpha_n < \beta_{n+1}$, then (7.34) has at least $2n$ nonnegative solutions $y_1, \dots, y_{2n} \in C[0,1]$ such that*

$$0 < \beta_1 < |y_1|_0 < \alpha_1 < \cdots < \alpha_n < |y_{2n}|_0 < \beta_{n+1}.$$

(II) *If $m = n$ and $0 < \beta_1 < \alpha_1 < \cdots < \beta_n < \alpha_n$, then (7.34) has at least $2n-1$ nonnegative solutions $y_1, \dots, y_{2n-1}$ such that*

$$0 < \beta_1 < |y_1|_0 < \alpha_1 < \cdots < \beta_n < |y_{2n-1}|_0 < \alpha_n.$$

(III) *If $n = m+1$ and $0 < \alpha_1 < \beta_1 < \cdots < \alpha_m < \beta_m < \alpha_{m+1}$, then (7.34) has at least $2m+1$ nonnegative solutions $y_0, \dots, y_{2m} \in C[0,1]$ such that*

$$0 \le |y_0|_0 < \alpha_1 < |y_1|_0 < \beta_1 < \cdots < \beta_m < |y_{2m}|_0 < \alpha_{m+1}.$$

(IV) *If $m = n$ and $0 < \alpha_1 < \beta_1 < \cdots < \alpha_n < \beta_n$, then (7.34) has at least $2n$ nonnegative solutions $y_0, \dots, y_{2n-1} \in C[0,1]$ such that*

$$0 \le |y_0|_0 < \alpha_1 < |y_1|_0 < \beta_1 < \cdots < \alpha_n < |y_{2n-1}|_0 < \beta_n.$$

Proof In (III) and (IV), the existence of a solution $y_0 \in C[0,1]$ of (7.34) such that $0 \le |y_0|_0 < \alpha_1$ follows from Theorem 7.11. The remainder of the proof follows by repeated application of Theorem 7.10. □

Remark 7.4 Similarly to Remark 7.1, we note that (7.53) may be replaced with

$$(7.54)\qquad \begin{cases} \text{there exist constants } \beta_j > 0,\ j = 1,\dots,m, \\ \text{for some } m \in \mathbf{N}, \text{ such that for each } j \in \{1,\dots,m\}, \\ (7.47) \text{ is satisfied with } \beta = \beta_j. \end{cases}$$

Suppose now that $|h|_0 = 0$. We give an example of a function f that satisfies conditions (7.41) and (7.47) (and hence (7.52) and (7.54) for various n and m respectively).

Example 7.2 Let $f(y) = y^\gamma$, $0 \leq \gamma < 1$. Then f satisfies (7.41) and (7.47) with

$$\alpha > K_1^{\frac{1}{1-\gamma}} \text{ and } \beta < (K_2 M^\gamma)^{\frac{1}{1-\gamma}}$$

respectively. Note that $\beta < \alpha$ since $K_2 \leq K_1$ and $0 < M < 1$.

Example 7.3 Suppose that $f(y) = y^\delta$, $\delta > 1$. Then f satisfies (7.41) and (7.47) with

$$\alpha < K_1^{\frac{1}{1-\delta}} \text{ and } \beta > \left(K_2 M^\delta\right)^{\frac{1}{1-\delta}}$$

respectively. In this case $\alpha < \beta$ since $\dfrac{1}{1-\delta} < 0$.

Example 7.4 Let

$$f(y) = 1 + y^\gamma + y^\beta,\ 0 \leq \gamma < 1 < \delta.$$

Since

$$\frac{y}{K_2\left(1 + (My)^\gamma + (My)^\delta\right)} \to 0 \text{ as } y \to 0^+ \text{ and } y \to \infty,$$

there exist $0 < \tilde{\beta}_1 < \tilde{\beta}_2$ such that f satisfies (7.47) with both $\beta = \beta_1$ and $\beta = \beta_2$, where $\beta_1 \in (0, \tilde{\beta}_1)$ and $\beta_2 \in (\tilde{\beta}_2, \infty)$. In addition, if

$$\sup_{y \in [0,\infty)} \frac{y}{K_1(1 + y^\gamma + y^\delta)} > 1,$$

then there exists $\alpha > 0$ for which f satisfies (7.41). Note that we can choose α such that $0 < \beta_1 < \alpha < \beta_2$.

Notes The fixed point theorems in Chapter 7 were adapted from Agarwal and O'Regan [2], and Simon and Volkmann [170]. Theorems 7.10–7.12 were taken from Meehan and O'Regan [125].

Exercises

7.1 Let E be a Banach space and $C \subseteq E$ a closed, convex, nonempty set with $\alpha u + \beta v \in C$ for all $\alpha \geq 0$, $\beta \geq 0$ and u, $v \in C$. In addition, let r, R be constants with $0 < r < R$. Suppose that $F : \overline{B}_R \to C$

is a continuous, k-set contractive map (here $0 \le k < 1$) and assume the following conditions hold:

(i) $x \neq \lambda F(x)$ for $\lambda \in [0,1)$ and $x \in S_R$,

(ii) there exists a $v \in C\backslash\{0\}$ with $x \neq F(x) + \delta v$ for any $\delta > 0$ and $x \in S_r$.

Show that F has a fixed point in $\{x : x \in C \text{ and } r \le \|x\| \le R\}$.

7.2 Let E, C, r and R be as in Exercise 7.1. Suppose that $F : \overline{B}_r \to C$ is a continuous, k-set contractive map (here $0 \le k < 1$), and assume the following conditions hold:

(iii) $x \neq \lambda F(x)$ for $\lambda \in [0,1)$ and $x \in S_r$,

(iv) there exists a $v \in C\backslash\{0\}$ with $x \neq F(x) + \delta v$ for any $\delta > 0$ and $x \in S_R$.

Show that F has a fixed point in $\{x : x \in C \text{ and } r \le \|x\| \le R\}$.

7.3 Let $E = (E, \|\cdot\|)$ be a Banach space, $C \subseteq E$ a cone and let $\|\cdot\|$ be increasing with respect to C. In addition, let r, R be constants with $0 < r < R$. Suppose that $F : \overline{\Omega}_R \cap C \to C$ is a continuous, k-set contractive map (here $0 \le k < 1$) and assume *either*

(v) $\|F(x)\| \le \|x\|$ for $x \in \partial_E \Omega_R \cap C$ and $\|F(x)\| > \|x\|$ for $x \in \partial_E \Omega_r \cap C$

or

(vi) $\|F(x)\| > \|x\|$ for $x \in \partial_E \Omega_R \cap C$ and $\|F(x)\| \le \|x\|$ for $x \in \partial_E \Omega_r \cap C$

holds.

Show that F has a fixed point in $C \cap \{x \in E : r \le \|x\| \le R\}$.

7.4 Let E, C, r and R be as in Exercise 7.3. Suppose that $F : \overline{\Omega}_R \cap C \to C$ is a continuous, 1-set contractive, demicompact map and assume there exists a $\delta > 0$ with *either*

(vii) $\|F(x)\| \le \|x\|$ for $x \in \partial_E \Omega_R \cap C$ and $\|F(x)\| \ge (1+\delta)\|x\|$ for $x \in \partial_E \Omega_r \cap C$

or

(viii) $\|F(x)\| \ge (1+\delta)\|x\|$ for $x \in \partial_E \Omega_R \cap C$ and $\|F(x)\| \le \|x\|$ for $x \in \partial_E \Omega_r \cap C$

holding.

Show that F has a fixed point in $C \cap \{x \in E : r \le \|x\| \le R\}$. (Recall that F is demicompact if each sequence $\{x_n\} \subseteq \overline{\Omega}_R \cap C$ has a convergent subsequence whenever $\{x_n - F(x_n)\}$ is a convergent sequence in E.)

8

Fixed Point Theory in Hausdorff Locally Convex Linear Topological Spaces

This chapter presents fixed point results for maps defined on Hausdorff locally convex linear topological spaces. We begin with the Schauder–Tychonoff theorem which is an extension of Schauder's fixed point theorem, which in turn is an extension of Brouwer's fixed point theorem. In the proof we will need the following approximation theorem.

Theorem 8.1 *Let E be a Hausdorff locally convex linear topological space, A a compact subset of E and C a convex subset of E with $A \subseteq C$. Then given an open neighbourhood U of 0 (the zero element of E), there exists a continuous mapping $x \mapsto P_U(x)$, from A into E, with*

(i) $P_U(x) \in L \cap C$ *for* $x \in A$
and
(ii) $P_U(x) - x \in U$ *for* $x \in A$;

here L is a finite dimensional subspace of E.

Proof Without loss of generality assume that U is convex and balanced. Let

$$|x|_U := \inf\{\alpha > 0 : x \in \alpha U\}$$

be the Minkowski functional associated with U. Of course $x \mapsto |x|_U$ is a continuous seminorm on E and

$$U = \{x : x \in E \text{ and } |x|_U < 1\}.$$

Since A is compact there exists a finite set $\{a_1, \ldots, a_n\} \subseteq A$ such that

$$A \subseteq \bigcup_{i=1}^{n} U(a_i)$$

where $U(a) := U + a$ for $a \in E$. Define the function μ_i, $i = 1, \ldots, n$, by

$$\mu_i(x) := \max\{0, 1 - |x - a_i|_U\} \text{ for } x \in E.$$

Since $|\cdot|_U$ is a continuous function on E we have that $\mu_i(\cdot)$, $i = 1, \ldots, n$, is also a continuous function on E. In addition for $i = 1, \ldots, n$,

$$0 \leq \mu_i(x) \leq 1 \text{ for } x \in E$$

with

$$\mu_i(x) = 0 \text{ if } x \notin U(a_i) \text{ and } \mu_i(x) > 0 \text{ otherwise.}$$

Let

$$P_U(x) := \frac{\sum_{i=1}^{n} \mu_i(x) a_i}{\sum_{i=1}^{n} \mu_i(x)} \text{ for } x \in A.$$

Notice that P_U is well defined since if $x \in A$ then $x \in U(a_i)$ for some $i \in \{1, \ldots, n\}$ and therefore $\sum_{i=1}^{n} \mu_i(x) \neq 0$. Note also that P_U is a continuous function on A. Its values of course belong to the linear subspace L, generated by $\{a_i : i = 1, \ldots, n\}$. In addition since $A \subseteq C$ and C is convex, we have that

$$P_U(x) \in C \text{ for each } x \in A.$$

Therefore

$$P_U(x) \in L \cap C \text{ for } x \in A.$$

Next notice that

$$P_U(x) - x = \frac{\sum_{i=1}^{n} \mu_i(x)(a_i - x)}{\sum_{i=1}^{n} \mu_i(x)} \text{ for } x \in A$$

and therefore

$$|P_U(x) - x|_U \leq \frac{\sum_{i=1}^{n} \mu_i(x)|a_i - x|_U}{\sum_{i=1}^{n} \mu_i(x)} < 1 \text{ for } x \in A$$

since for any $i = 1, \ldots, n$, either $\mu_i(x) = 0$ and $|a_i - x|_U \geq 1$, or $\mu_i(x) > 0$ and $|a_i - x|_U < 1$. This immediately yields $P_U(x) - x \in U$ for $x \in A$. □

Our next result is known as the Schauder–Tychonoff theorem.

Theorem 8.2 *Let E be a Hausdorff locally convex linear topological space, C a convex subset of E and $F : C \to E$ a continuous mapping such that*

$$F(C) \subseteq A \subseteq C$$

with A compact. Then F has at least one fixed point.

Proof Let U be an open, convex, balanced neighbourhood of 0 (the zero element of E) and P_U be as in Theorem 8.1. Define a function F_U by

$$F_U(x) := P_U(F(x)) \text{ for } x \in C.$$

Since P_U takes values in the space L (defined in Theorem 8.1), we shall restrict our considerations to this space. From Theorem 8.1 we have that

$$F_U(L \cap C) \subseteq P_U(A) \subseteq L \cap C \tag{8.1}$$

since if $x \in L \cap C$ then $F(x) \in A$ and therefore

$$F_U(x) = P_U(F(x)) \subseteq L \cap C.$$

Let $K^\star$ denote the convex hull of the compact set $P_U(A)$ in L. Note that $K^\star$ is compact. In addition (8.1) and

$$P_U(A) \subseteq K^\star \subseteq L \cap C$$

implies that

$$F_U(K^\star) \subseteq K^\star. \tag{8.2}$$

Apply Brouwer's fixed point theorem to deduce that there exists $x \in K^\star$ with $x = F_U(x)$. That is, x satisfies

$$x - F(x) \in U \tag{8.3}$$

since $x = F_U(x)$ is equivalent to $x = P_U(F(x))$, and therefore by Theorem 8.1 we have that

$$P_U(F(x)) - F(x) \in U.$$

We have shown:

$$(8.4) \qquad \begin{cases} \text{to any open neighbourhood } U \text{ of } 0, \text{ there exists} \\ \text{at least one } x \in K^\star \subseteq C \text{ such that (8.3) holds.} \end{cases}$$

Suppose now that $x \neq F(x)$ for all $x \in C$. The continuity of F and the fact that E is Hausdorff guarantee that there exist two open neighbourhoods V_x and W_x of 0 with the properties

$$(8.5) \qquad F(C \cap V_x(x)) \subseteq W_x(F(x))$$

and

$$(8.6) \qquad V_x(x) \cap W_x(F(x)) \neq \varnothing.$$

Choose U_x to be another open neighbourhood of 0 such that

$$(8.7) \qquad 2U_x \subseteq V_x \cap W_x.$$

Since A is compact there exists a finite set $\{a_i : i = 1, \dots, n\} \subseteq A$ with

$$A \subseteq \bigcup_{i=1}^{n} U_{a_i}(a_i).$$

We *claim* that for any $x \in C$ there exists $j \in \{1, \dots, n\}$ such that

$$(8.8) \qquad x - F(x) \subseteq U_{a_j}$$

cannot hold. Fix $x \in C$. Since $y = F(x) \in A$ there exists $j \in \{1, \dots, n\}$ with $y \in U_{a_j}(a_j)$. In addition we have

$$(8.9) \qquad U_{a_j}(y) \subseteq V_{a_j}(a_j).$$

To see this notice that

$$y = u + a_j \text{ for some } u \in U_{a_j},$$

therefore if $z \in U_{a_j}(y)$ there exists $w \in U_{a_j}$ with

$$z = w + y = w + u + a_j,$$

and consequently

$$z \in 2U_{a_j} + a_j \subseteq V_{a_j}(a_j)$$

from (8.7).

Suppose that (8.8) is not true. Then for any $x \in C$ we have that $x \in U_{a_j}(y)$ with $y = F(x)$, and therefore from (8.9), we see that $x \in V_{a_j}(a_j)$. Now (8.5) guarantees that

$$y = F(x) \in W_{a_j}(F(a_j)).$$

However, $y \in W_{a_j}(F(a_j))$ and (8.6) imply

$$y \notin V_{a_j}(a_j)$$

which contradicts (8.9). Therefore (8.8) cannot be true.

Choose U such that

$$U \subseteq \bigcap_{i=1}^{n} U_{a_i}.$$

From what we deduced above, it follows that

$$x - F(x) \notin U \text{ for all } x \in C.$$

This however contradicts (8.4). Consequently there exists $x \in C$ with $x = F(x)$. □

Theorem 8.3 *Let C be a convex subset of a Hausdorff locally convex linear topological space E. Suppose that $F : C \to C$ is a continuous, compact map. Then F has at least one fixed point in C.*

Proof Apply Theorem 8.2 recalling that $F : C \to C$ is compact if $F(C)$ is a relatively compact subset of C. □

Next we establish a nonlinear alternative of Leray–Schauder type for maps defined on Hausdorff locally convex linear topological spaces.

Theorem 8.4 *Let E be a Hausdorff locally convex linear topological space, C a convex subset of E, U an open subset of C and $p \in U$. Suppose that $F : \overline{U} \to C$ (here $\overline{U}$ denotes the closure of U in C) is a continuous, compact map. Then either*

(A1) *F has a fixed point in $\overline{U}$, or*

(A2) *there are a $u \in \partial U$ (the boundary of U in C) and $\lambda \in (0,1)$ with $u = \lambda F(u) + (1-\lambda)p$.*

Proof Suppose that (A2) does not hold and F has no fixed points on ∂U (otherwise we are finished). Consider

$$A := \{x \in \overline{U} : x = tF(x) + (1-t)p \text{ for some } t \in [0,1]\}.$$

Now $A \neq \varnothing$ since $p \in U$. To see that A is closed (in C), let (x_α) be a net in A (that is, $x_\alpha = t_\alpha F(x_\alpha) + (1-t_\alpha)p$ for some $t_\alpha \in [0,1]$) with $x_\alpha \to x \in \overline{U}$. Without loss of generality assume that $t_\alpha \to t \in [0,1]$. Let $R : \overline{U} \times [0,1] \to C$ be given by

$$R(x,t) := tF(x) + (1-t)p.$$

The continuity of R together with

$$(x_\alpha, t_\alpha) \to (x, t) \text{ and } x_\alpha = R(x_\alpha, t_\alpha)$$

guarantees that

$$x = R(x, t) := tF(x) + (1 - t)p.$$

Thus A is closed. In addition $F : \overline{U} \to C$ being a compact map implies that

$$\overline{A}\ (= A) \text{ is compact.}$$

Notice also that $A \cap \partial U = \varnothing$. Since C is completely regular, A is compact and ∂U is closed, there exists a continuous function

$$\mu : \overline{U} \to [0, 1] \text{ with } \mu(A) = 1 \text{ and } \mu(\partial U) = 0.$$

Let

$$N(x) := \begin{cases} \mu(x)F(x) + (1 - \mu(x))p, & x \in \overline{U}, \\ p, & x \in C \backslash \overline{U}. \end{cases}$$

It is immediate that $N : C \to C$ is a continuous, compact map since $F : \overline{U} \to C$ is compact. Theorem 8.3 guarantees the existence of an $x \in C$ with $x = N(x)$. Notice that $x \in U$ since $p \in U$. As a result $x = \mu(x)F(x) + (1 - \mu(x))p$, therefore $x \in A$ and we have that $\mu(x) = 1$. This implies that $x = F(x)$. □

We now use Theorem 8.4 to derive a fixed point theorem of Furi–Pera type. This will be particularly useful in applications as we will demonstrate after the proof of the theorem.

Theorem 8.5 *Let E be a metrisable locally convex linear topological space, Q a closed, convex subset of E and $0 \in Q$. Suppose that $F : Q \to E$ is a continuous, compact map and assume the following condition holds:*

$$(8.10) \qquad \begin{cases} \textit{if } \{(x_j, \lambda_j)\}_{j=1}^{\infty} \textit{ is a sequence in } \partial Q \times [0, 1] \\ \textit{converging to } (x, \lambda) \textit{ with } x = \lambda F(x) \textit{ and } 0 \le \lambda < 1, \\ \textit{then } \lambda_j F(x_j) \in Q \textit{ for } j \textit{ sufficiently large.} \end{cases}$$

Then F has a fixed point in Q.

Proof Let $r : E \to Q$ be a continuous retraction (the existence of r follows from Dugundji's extension theorem).

Remark 8.1 If $0 \in \operatorname{int} Q$ we may take

$$r(x) := \frac{x}{\max\{1, \mu(x)\}} \text{ for } x \in E,$$

where μ is the Minkowski functional on Q, that is,

$$\mu(x) := \inf\{\alpha > 0 : x \in \alpha Q\}.$$

Note that if $\operatorname{int} Q = \varnothing$ then $\partial Q = Q$.

From Remark 8.1 we may choose (and we do so) the retraction r as above so that

$$r(z) \in \partial Q \text{ for } z \in E \backslash Q.$$

Consider

$$B := \{x \in E : x = Fr(x)\}.$$

Firstly $B \neq \varnothing$. To see this notice that since r is continuous, $Fr : E \to E$ is a continuous, compact map. Theorem 8.3 guarantees that Fr has a fixed point and therefore $B \neq \varnothing$. In addition B is closed. To see this let (x_α) be a net in B with $x_\alpha \to x \in E$. By the continuity of Fr it follows that $x = Fr(x)$ and therefore $x \in B$. In fact B is compact since $F : Q \to E$ is a compact map and

$$B \subseteq Fr(B) \subseteq F(Q).$$

It remains to show that $B \cap Q \neq \varnothing$. To do this we argue by contradiction. Suppose that $B \cap Q = \varnothing$. Then since B is compact and Q is closed, there exists a $\delta > 0$ with $\operatorname{dist}(B, Q) > \delta$. Choose $m \in \{1, 2, \ldots\}$ with $1 < \delta m$. Define

$$U_i := \left\{x \in E : d(x, Q) < \frac{1}{i}\right\} \text{ for } i \in \{m, m+1, \ldots\};$$

here d is the metric associated with E. Fix $i \in \{m, m+1, \ldots\}$. Since $\operatorname{dist}(B, Q) > \delta$ then we see that $B \cap \overline{U_i} = \varnothing$. In addition U_i is open, $0 \in U_i$ and $Fr : \overline{U_i} \to E$ is a continuous, compact map. Theorem 8.4 guarantees, since $B \cap \overline{U_i} = \varnothing$, that there exists

$$(y_i, \lambda_i) \in \partial U_i \times (0, 1) \text{ with } y_i = \lambda_i Fr(y_i).$$

Consequently

$$\lambda_j Fr(y_j) \notin Q \text{ for } j \in \{m, m+1, \ldots\}. \tag{8.11}$$

We now look at

$$D := \{x \in E : x = \lambda Fr(x) \text{ for some } \lambda \in [0, 1]\}.$$

Notice that $D \neq \varnothing$ is closed and in fact compact since $F : Q \to E$ is compact. This together with

$$d(y_j, Q) = \frac{1}{j} \text{ and } |\lambda_j| \leq 1 \text{ for } j \in \{m, m+1, \ldots\}$$

implies that we may assume without loss of generality that

$$\lambda_j \to \lambda^\star \in [0,1] \text{ and } y_j \to y^\star \in \partial Q.$$

In addition we have

$$y_j = \lambda_j Fr(y_j) \to \lambda^\star Fr(y^\star)$$

and therefore $y^\star = \lambda^\star Fr(y^\star)$. Note that $\lambda^\star \neq 1$ since $B \cap Q = \varnothing$. Hence $0 \leq \lambda^\star < 1$. However, (8.10) with

$$x_j = r(y_j) \in \partial Q \text{ and } x = y^\star = r(y^\star)$$

implies that $\lambda_j Fr(y_j) \in Q$ for j sufficiently large. This contradicts (8.11). Thus $B \cap Q \neq \varnothing$, therefore there exists $x \in Q$ with $x = Fr(x) = F(x)$. □

We now illustrate how Theorem 8.5 can be applied in practice. In particular we will establish existence principles and results for the second order boundary value problems

$$\text{(8.12)} \qquad \begin{cases} y'' + m^2 y' = f(t, y), \text{ a.e. on } [0, \infty), \\ y(0) = a, \ \lim_{t \to \infty} y(t) = 0 \end{cases}$$

and

$$\text{(8.13)} \qquad \begin{cases} y'' - m^2 y = f(t, y), \text{ a.e. on } [0, \infty), \\ y(0) = a, \ \lim_{t \to \infty} y(t) = 0; \end{cases}$$

here $m \neq 1$ is a constant and y takes values in $\mathbf{R}$. Let $C([0,\infty), \mathbf{R})$ be the space of continuous mappings from $[0, \infty)$ to $\mathbf{R}$, the topology being that of uniform convergence on each compact interval of $[0, \infty)$. If $u \in C([0, \infty), \mathbf{R})$ then for every $m \in \{1, 2, \ldots\}$, we define the seminorms $\rho_m(u)$ by

$$\rho_m(u) := \sup_{t \in [0, t_m]} |u(t)|$$

where $t_m \uparrow \infty$; the metric (since the topology is determined by a countable number of seminorms) is defined by

$$d(x, y) := \sum_{m=1}^{\infty} \frac{1}{2^m} \frac{\rho_m(x - y)}{1 + \rho_m(x - y)}.$$

Note that $C([0,\infty),\mathbf{R})$ is a Fréchet space (a complete metrisable locally convex linear topological space). $BC[0,\infty)$ will denote the space of bounded continuous mappings from $[0,\infty)$ to $\mathbf{R}$. If $u \in BC[0,\infty)$ then we write

$$|u|_\infty := \sup_{t\in[0,\infty)} |u(t)|.$$

The Arzelà–Ascoli theorem says that a set $\Omega \subseteq C([0,\infty),\mathbf{R})$ is relatively compact if Ω is uniformly bounded and equicontinuous on each compact interval of $[0,\infty)$.

Theorem 8.6 *Assume that the following conditions are satisfied:*

$$(8.14)\quad \begin{cases} f:[0,\infty)\times\mathbf{R}\to\mathbf{R} \text{ is an } L\text{-Carathéodory function;} \\ \text{(i)}\quad s\mapsto f(s,y) \text{ is measurable for any } y\in\mathbf{R}, \\ \text{(ii)}\quad y\mapsto f(s,y) \text{ is continuous for a.e. } s\in[0,\infty), \\ \text{(iii) for each } r>0 \text{ there exists } \tau_r\in L^1[0,\infty) \text{ such that} \\ \qquad |y|\le r \text{ implies } |f(s,y)|\le \tau_r(s) \text{ for almost all} \\ \qquad s\in[0,\infty) \text{ and } \lim_{t\to\infty} e^{-m^2t}\int_0^t e^{m^2s}\tau_r(s)\,ds=0; \end{cases}$$

and

$$(8.15)\quad \begin{cases} \text{there exists a constant } M_0>|a| \text{ with } |u(t)|\le M_0,\ t\in[0,\infty), \\ \text{for any function } u\in BC([0,\infty),\mathbf{R})\cap W^{2,1}_{loc}([0,\infty),\mathbf{R}) \\ \text{which satisfies } u''+m^2u'=\lambda f(t,u) \text{ a.e. on } [0,\infty), \\ u(0)=a,\ \lim_{t\to\infty}u(t)=0 \text{ for } 0\le\lambda<1. \end{cases}$$

Then (8.12) *has a solution* $y\in BC([0,\infty),\mathbf{R})\cap W^{2,1}_{loc}([0,\infty),\mathbf{R})$.

Remark 8.2

(i) Let $I=[a,b]$. The Sobolev class of functions u such that $u^{(m-1)}$ is absolutely continuous and $u^{(m)}\in L^p(I)$ is denoted by $W^{m,p}(I,\mathbf{R})$.

(ii) By a solution of (8.12) we mean a function

$$y\in BC([0,\infty),\mathbf{R})\cap W^{2,1}_{loc}([0,\infty),\mathbf{R})$$

which satisfies the differential equation almost everywhere on $[0,\infty)$ and the stated boundary data.

(iii) If in addition $f:[0,\infty)\times\mathbf{R}\to\mathbf{R}$ is continuous, then by a solution of (8.12) we mean a function $y\in C^2([0,\infty),\mathbf{R})$ which satisfies the differential equation everywhere on $[0,\infty)$ and the stated boundary data. Also, in assumption (8.15), in this case $u \in C^2([0,\infty),\mathbf{R})$ and u satisfies the differential equation everywhere.

Proof of Theorem 8.6 It is easy to see from (8.14) that solving (8.12) is equivalent to finding a $y \in BC([0,\infty),\mathbf{R})$ which satisfies

$$(8.16) \qquad y(t) = ae^{-m^2t} - \frac{1}{m^2}\int_t^\infty f(s,y(s))\,ds - \frac{e^{-m^2t}}{m^2}\int_0^t e^{m^2s}f(s,y(s))\,ds + \frac{e^{-m^2t}}{m^2}\int_0^\infty f(s,y(s))\,ds.$$

Let $E := C([0,\infty),\mathbf{R})$,

$$Q := \{y \in C([0,\infty),\mathbf{R}) : y \in BC([0,\infty),\mathbf{R}) \text{ with } |y|_\infty \le M_0 + 1 \equiv N_0\}$$

and

$$Fy(t) := ae^{-m^2t} - \frac{1}{m^2}\int_t^\infty f(s,y(s))\,ds - \frac{e^{-m^2t}}{m^2}\int_0^t e^{m^2s}f(s,y(s))\,ds + \frac{e^{-m^2t}}{m^2}\int_0^\infty f(s,y(s))\,ds.$$

Now Q is a closed, convex, bounded subset of $C([0,\infty),\mathbf{R})$. We now show that $F : Q \to C([0,\infty),\mathbf{R})$ is a continuous, compact map. First we examine continuity. Let $y_n \to y$ in Q. Then there exists $\tau_{N_0} \in L^1[0,\infty)$ with

$$|f(s,y_n(s))| \le \tau_{N_0}(s) \text{ and } |f(s,y(s))| \le \tau_{N_0}(s) \text{ for almost all } s \in [0,\infty).$$

In addition for each $t \in [0,\infty)$ we have

$$f(s,y_n(s)) \to f(s,y(s)) \text{ for a.a. } s \in [0,\infty).$$

This together with the Lebesgue dominated convergence theorem implies that $Fy_n(s) \to Fy(s)$ pointwise on $[0,t_m]$. Let $x,t \in [0,t_m]$ with $t < x$. Then

$$|Fy_n(t) - Fy_n(x)| \le |a|\left|e^{-m^2t} - e^{-m^2x}\right| + \frac{1}{m^2}\int_t^x \tau_{N_0}(s)\,ds + \frac{1}{m^2}\left[e^{-m^2t} - e^{-m^2x}\right]\int_0^t \tau_{N_0}(s)\,ds + \frac{e^{-m^2t}}{m^2}\int_t^x \tau_{N_0}(s)\,ds$$

$$+\frac{1}{m^2}\left[e^{-m^2t}-e^{-m^2x}\right]\int_0^\infty \tau_{N_0}(s)\,ds.$$

A similar bound can be found for $|Fy(t) - Fy(x)|$. Therefore, given $\epsilon > 0$, there exists a $\delta > 0$ such that for any $t, x \in [0, t_m]$ and $|t - x| < \delta$ we have

$$|Fy_n(t) - Fy_n(x)| < \epsilon \text{ for all } n \tag{8.17}$$

and

$$|Fy(t) - Fy(x)| < \epsilon. \tag{8.18}$$

Hence (8.17), (8.18) and the fact that $Fy_n(s) \to Fy(s)$ pointwise on $[0, t_m]$ imply the convergence is uniform on $[0, t_m]$. As a result $F : Q \to E$ is continuous.

We next show that $F(Q)$ is relatively compact in E. This follows once we illustrate the uniform boundedness and relative compactness of $F(Q)$ on $[0, t_m]$. We know that there exists $\tau_{N_0} \in L^1[0, \infty)$ with $|f(s, u)| \le \tau_{N_0}$ for almost every $s \in [0, \infty)$ and $|u| \le M_0$. The equicontinuity of $F(Q)$ on $[0, t_m]$ follows essentially by the same reasoning as that used to prove (8.18). In addition, $F(Q)$ is uniformly bounded since for $t \in [0, t_m]$ we have that

$$|Fu(t)| \le |a| + \frac{3}{m^2}\int_0^\infty \tau_{N_0}(s)\,ds$$

for each $u \in Q$. Thus $F(Q)$ is relatively compact in E and hence $F : Q \to E$ is a compact map.

The result follows immediately from Theorem 8.5 once we show that (8.10) is satisfied. Take a sequence $\{(y_j, \lambda_j)\}_{j=1}^\infty$ in $\partial Q \times [0, 1]$ with $\lambda_j \to \lambda$ and $y_j \to y$ with $y = \lambda F(y)$ and $0 \le \lambda < 1$. We need to show that $\lambda_j F(y_j) \in Q$ for j sufficiently large. Take any $v \in E$ and $|v(t)| \le N_0$ for $t \in [0, \infty)$. Then

$$\begin{aligned} |Fv(t)| &\le |a|e^{-m^2t} + \frac{1}{m^2}\int_t^\infty \tau_{N_0}(s)\,ds + e^{-m^2t}\int_0^t e^{m^2s}\tau_{N_0}(s)\,ds \\ &\quad + \frac{e^{-m^2t}}{m^2}\int_0^\infty \tau_{N_0}(s)\,ds \equiv \Psi_{N_0}(t). \end{aligned}$$

Notice that $\lim_{t\to\infty} \Psi_{N_0}(t) = 0$. This together with the fact that $y_j \in Q$ implies that there exists an $a_0 \ge 0$ with

$$|Fy_j(t)| \le M_0 + 1 \equiv N_0, \text{ for } t \in [a_0, \infty) \text{ and } j \in \{1, 2, \dots\}.$$

Consequently

$$(8.19) \qquad |\lambda_j F y_j(t)| \leq N_0 \text{ for } t \in [a_0, \infty) \text{ and } j \in \{1, 2, \ldots\}.$$

Next consider the situation where $t \in [0, a_0]$. Since F is continuous on Q we have that $Fy_j \to Fy$ uniformly on $[0, a_0]$. In addition since $\lambda_j \to \lambda$ and $F(Q)$ is a bounded set in E we see that

$$\lambda_j F y_j \to \lambda F y \text{ uniformly on } [0, a_0].$$

Thus there exists $j_0 \in \{1, 2, \ldots\}$ with

$$(8.20) \qquad |\lambda_j F y_j(t)| \leq |\lambda F y(t)| + 1, \ t \in [0, a_0], \text{ for } j \geq j_0.$$

Now $y = \lambda F(y)$; therefore (8.15) implies for $j \geq j_0$ that

$$(8.21) \qquad |\lambda_j F y_j(t)| \leq M_0 + 1 = N_0 \text{ for } t \in [0, a_0].$$

Now (8.19) and (8.21) imply that $\lambda_j F(y_j) \in Q$ for $j \geq j_0$. Consequently all the conditions of Theorem 8.5 are satisfied and therefore (8.12) has a solution. □

We also have an existence principle for (8.13).

Theorem 8.7 *Assume that the following conditions are satisfied:*

$$(8.22) \quad \begin{cases} f : [0,\infty) \times \mathbf{R} \to \mathbf{R} \text{ is an } LL\text{-Carathéodory function;} \\ \text{(i)} \quad s \mapsto f(s,y) \text{ is measurable for any } y \in \mathbf{R}, \\ \text{(ii)} \quad y \mapsto f(s,y) \text{ is continuous for a.e. } s \in [0,\infty), \\ \text{(iii) for each } r > 0 \text{ there exists } \tau_r \in L^1[0,\infty) \text{ such that} \\ \qquad |y| \leq r \text{ implies } |f(s,y)| \leq \tau_r(s) \text{ for almost all} \\ \qquad s \in [0,\infty) \text{ together with } \lim_{t\to\infty} e^{-mt} \int_0^t e^{ms} \tau_r(s)\, ds = 0 \\ \qquad \text{and } \lim_{t\to\infty} e^{mt} \int_t^\infty e^{-ms} \tau_r(s)\, ds = 0; \end{cases}$$

and

$$(8.23) \quad \begin{cases} \text{there exists a constant } M_0 > |a| \text{ with } |u(t)| \leq M_0,\ t \in [0,\infty), \\ \text{for any function } u \in BC([0,\infty),\mathbf{R}) \cap W^{2,1}_{loc}([0,\infty),\mathbf{R}) \\ \text{which satisfies } u'' - m^2 u = \lambda f(t,u) \text{ a.e. on } [0,\infty), \\ u(0) = a,\ \lim_{t\to\infty} u(t) = 0 \text{ for } 0 \leq \lambda < 1. \end{cases}$$

Then (8.13) has a solution $y \in BC([0,\infty),\mathbf{R}) \cap W^{2,1}_{loc}([0,\infty),\mathbf{R})$.

Proof If follows from (8.22) that solving (8.13) is equivalent to finding a $y \in BC([0,\infty),\mathbf{R})$ which satisfies

$$\begin{aligned} y(t) &= ae^{-mt} + \frac{e^{-mt}}{2m}\int_0^\infty e^{-ms} f(s,y(s))\,ds \\ &\quad -\frac{e^{mt}}{2m}\int_t^\infty e^{-ms} f(s,y(s))\,ds - \frac{e^{-mt}}{2m}\int_0^t e^{ms} f(s,y(s))\,ds. \end{aligned}$$

Let

$$Q := \{y \in C([0,\infty),\mathbf{R}) : y \in BC([0,\infty),\mathbf{R}) \text{ with } |y|_\infty \le M_0 + 1 \equiv N_0\}$$

and

$$\begin{aligned} Fy(t) &:= ae^{-mt} + \frac{e^{-mt}}{2m}\int_0^\infty e^{-ms} f(s,y(s))\,ds \\ &\quad -\frac{e^{mt}}{2m}\int_t^\infty e^{-ms} f(s,y(s))\,ds - \frac{e^{-mt}}{2m}\int_0^t e^{ms} f(s,y(s))\,ds. \end{aligned}$$

Essentially the same reasoning as in Theorem 8.6 will now establish the result. □

The general existence principles derived above can now be easily used to establish existence results for (8.12) and (8.13).

Example 8.1 Consider the boundary value problem

$$(8.24) \qquad \begin{cases} y'' + m^2 y' = f(t,y), \; 0 \le t < \infty, \; m \ne 0, \\ y(0) = 0, \; \lim_{t\to\infty} y(t) = 0. \end{cases}$$

Assume that (8.14),

$$(8.25) \qquad f : [0,\infty) \times \mathbf{R} \to \mathbf{R} \text{ is continuous}$$

and

$$(8.26) \qquad \begin{cases} \text{there exists a constant } M_0 > 0 \text{ such that} \\ |y| > M_0 \text{ implies that } yf(t,y) > 0 \text{ for all } t \in [0,\infty) \end{cases}$$

hold. Then (8.12) has a solution $y \in BC([0,\infty),\mathbf{R}) \cap C^2([0,\infty),\mathbf{R})$.

This follows immediately from Theorem 8.6 once we show that (8.15) holds. To see this let $u \in BC([0,\infty),\mathbf{R}) \cap C^2([0,\infty),\mathbf{R})$ be a solution of

$$\begin{cases} u'' + m^2 u' = \lambda f(t,u), \; 0 \le t < \infty, \; 0 \le \lambda < 1, \\ u(0) = 0, \; \lim_{t\to\infty} u(t) = 0. \end{cases}$$

We claim that $|u(t)| \le M_0$ for $t \in [0,\infty)$. If $\lambda = 0$ this is true since

$u \equiv 0$. Therefore suppose that $0 < \lambda < 1$. If there exists a $t \in (0, \infty)$ with $|u(t)| > M_0$ then

$$\max_{t\in[0,\infty)} |u(t)| = |u(t_0)| > M_0$$

with $t_0 \in (0, \infty)$ and $u'(t_0) = 0$. Consequently

$$\begin{aligned} u(t_0)u''(t_0) &= u(t_0)[u''(t_0) + m^2 u'(t_0)] \\ &= \lambda u(t_0) f(t_0, u(t_0)) > 0, \end{aligned}$$

which contradicts the maximality of $|u(t_0)|$. Hence $|u(t)| \le M_0$ for $t \in [0, \infty)$ and therefore (8.15) holds. Existence of a solution of (8.24) is now guaranteed from Theorem 8.6.

Example 8.2 Consider the boundary value problem

$$(8.27) \qquad \begin{cases} y'' - m^2 y = f(t, y), \ 0 \le t < \infty, \ m \ne 0, \\ y(0) = 0, \ \lim_{t\to\infty} y(t) = 0. \end{cases}$$

Assume that (8.22), (8.25) and

$$(8.28) \qquad \begin{cases} \text{there exists a constant } M_0 > 0 \text{ such that} \\ |y| > M_0 \text{ implies that } yf(t, y) \ge 0 \text{ for all } t \in [0, \infty). \end{cases}$$

Then (8.13) has a solution $y \in BC([0, \infty), \mathbf{R}) \cap C^2([0, \infty), \mathbf{R})$.

The result follows immediately from Theorem 8.7. We need only check that (8.23) holds. Let $u \in BC([0, \infty), \mathbf{R}) \cap C^2([0, \infty), \mathbf{R})$ be a solution of

$$\begin{cases} u'' - m^2 u = \lambda f(t, u), \ 0 \le t < \infty, \ 0 \le \lambda < 1, \\ u(0) = 0, \ \lim_{t\to\infty} u(t) = 0. \end{cases}$$

Again assume that $0 < \lambda < 1$ and $\max_{t\in[0,\infty)} |u(t)| = |u(t_0)| > M_0$ with $t_0 \in (0, \infty)$. Then

$$u(t_0)u''(t_0) = m^2[u(t_0)]^2 + \lambda u(t_0) f(t_0, u(t_0)) > 0$$

– a contradiction.

Finally in this chapter we present a continuation principle for continuous, compact maps defined on Fréchet spaces. In particular we present the 0-epi map approach by Furi, Martelli and Vignoli. For the remainder of this chapter E will be a Fréchet space and U an open subset of E (in fact any normal locally convex linear topological space will suffice).

Definition 8.1 We let $M_{\partial U}(\overline{U}, E)$ denote the set of continuous maps $F : \overline{U} \to E$ with $F(x) \neq 0$ for $x \in \partial U$.

Definition 8.2 We let $K(\overline{U}, E)$ denote the set of continuous, compact maps $F : \overline{U} \to E$.

Definition 8.3 We let $OK_{\partial U}(\overline{U}, E)$ denote the maps $F \in K(\overline{U}, E)$ with $F(x) = 0$ for $x \in \partial U$.

Definition 8.4 A map $F \in M_{\partial U}(\overline{U}, E)$ is called 0-epi if for every $G \in OK_{\partial U}(\overline{U}, E)$ there exists $x \in U$ with $F(x) = G(x)$.

Remark 8.3 If $F \in M_{\partial U}(\overline{U}, E)$ is 0-epi then there exists $x \in U$ with $F(x) = 0$.

Theorem 8.8 *Let E be a Fréchet space, U an open subset of E and $0 \in U$. Then the identity map $i : \overline{U} \to E$, given by $i(x) := x$, is 0-epi.*

Proof Let $G \in OK_{\partial U}(\overline{U}, E)$. We must show that there exists $x \in U$ with $x = G(x)$. Define the map $J : E \to E$ by

$$J(x) := \begin{cases} G(x), & x \in \overline{U}, \\ 0, & x \in E \backslash \overline{U}. \end{cases}$$

Notice that $J : E \to E$ is a continuous, compact map. The Schauder–Tychonoff theorem guarantees that there exists $x \in E$ with $x = J(x)$. In fact $x \in U$ since $0 \in U$ and as a result $x = G(x)$. □

We next show that the property of being 0-epi is invariant by homotopy for compact maps.

Theorem 8.9 *Let E be a Fréchet space and U an open subset of E. Suppose that $F \in M_{\partial U}(\overline{U}, E)$ is 0-epi and $H : \overline{U} \times [0, 1] \to E$ is a continuous, compact map with $H(x, 0) = 0$ for every $x \in \partial U$. In addition assume that*

$$F(x) \neq H(x, t) \text{ for all } x \in \partial U \text{ and } t \in [0, 1] \tag{8.29}$$

holds. Then $F(\cdot) - H(\cdot, 1) : \overline{U} \to E$ is 0-epi.

Proof Let $G \in OK_{\partial U}(\overline{U}, E)$ and consider

$$B := \{x \in \overline{U} : F(x) = G(x) + H(x, t) \text{ for some } t \in [0, 1]\}.$$

When $t = 0$ we have

$$G(\cdot) + H(\cdot, 0) \in OK_{\partial U}(\overline{U}, E)$$

and this together with the fact that F is 0-epi implies that $B \neq \emptyset$. It is immediate that B is closed and (8.29) guarantees that $B \cap \partial U = \emptyset$. Thus there exists a continuous

$$\mu : \overline{U} \to [0, 1] \text{ with } \mu(\partial U) = 0 \text{ and } \mu(B) = 1.$$

Define a map $J : \overline{U} \to E$ by

$$J(x) := G(x) + H(x, \mu(x)).$$

Clearly J is a continuous, compact map. Also for $x \in \partial U$ we have that

$$J(x) = 0 + H(x, 0) = 0$$

and as a result $J \in OK_{\partial U}(\overline{U}, E)$. Now since F is 0-epi there exists $x \in U$ with $F(x) = G(x) + H(x, \mu(x))$. Thus $x \in B$ (by definition) and consequently $\mu(x) = 1$. Hence $F(x) = G(x) + H(x, 1)$ and we are finished. □

We can deduce some applicable results from Theorem 8.9.

Theorem 8.10 *Let E be a Fréchet space and U an open subset of E. Suppose that $F \in M_{\partial U}(\overline{U}, E)$ is 0-epi and $G \in K(\overline{U}, E)$. Then either*

(A1) *there exists $x \in \overline{U}$ with $F(x) = G(x)$, or*
(A2) *there exists $x \in \partial U$ and $\lambda \in (0, 1)$ with $F(x) = \lambda G(x)$.*

Proof Assume that (A2) does not hold and $F(x) \neq G(x)$ for $x \in \partial U$ (otherwise we are finished). Consequently

$$\text{there exist } x \in \partial U \text{ and } \lambda \in [0, 1] \text{ with } F(x) = \lambda G(x) \tag{8.30}$$

cannot occur. Let $H : \overline{U} \times [0, 1] \to E$ be defined by

$$H(x, t) := tG(x).$$

Clearly H is a continuous, compact map with $H(x, 0) = 0$ for $x \in \partial U$ (in fact for $x \in \overline{U}$). Now Theorem 8.9 implies that $F(\cdot) - G(\cdot)$ is 0-epi, therefore there exists $x \in U$ with $F(x) - G(x) = 0$, in other words, (A1) occurs. □

Theorem 8.11 *Let E be a Fréchet space, U an open subset of E and $0 \in U$. Suppose that $G \in K(\overline{U}, E)$. Then either*

(A1) *there exists* $x \in \overline{U}$ *with* $x = G(x)$, *or*

(A2) *there exist* $x \in \partial U$ *and* $\lambda \in (0,1)$ *with* $x = \lambda G(x)$.

Remark 8.4 Of course Theorem 8.11 is our nonlinear alternative of Leray–Schauder type in Fréchet spaces.

Proof of Theorem 8.11 Let $F = I$ (the identity map). Theorem 8.8 implies that $F \in M_{\partial U}(\overline{U}, E)$ is 0-epi. The result follows immediately from Theorem 8.11. □

Notes Theorems 8.5–8.7 are taken from O'Regan [139], while the 0-epi map approach is adapted from Furi, Martelli and Vignoli [70].

Exercises

8.1 Let E be a Banach space and $C \subseteq E$ a closed, convex set. Suppose that $F : C \to C$ is weakly continuous and $F(C)$ is relatively compact in the weak topology. Show that F has a fixed point.

8.2 Let E be a Hausdorff locally convex linear topological space, U a convex, symmetric neighbourhood of the origin in E, $F : \overline{U} \to E$ a continuous, compact map and $F(-x) = -F(x)$ for all $x \in \partial U$. Show that F has a fixed point.

8.3 Let E be a Hausdorff locally convex linear topological space, C a closed, convex subset of E and U an open subset of C. As in Chapter 6, $K(\overline{U}, C)$ denotes the set of all continuous, compact maps $F : \overline{U} \to C$, with $K_{\partial U}(\overline{U}, C)$ denoting the maps $F \in K(\overline{U}, C)$ with $x \neq F(x)$ for $x \in \partial U$. A map $F \in K_{\partial U}(\overline{U}, C)$ is essential in $K_{\partial U}(\overline{U}, C)$ if for every map $H \in K_{\partial U}(\overline{U}, C)$ with $H|_{\partial U} = F|_{\partial U}$ there exists $x \in U$ with $x = H(x)$.

(a) Suppose that $F, G \in K_{\partial U}(\overline{U}, C)$ with $F \simeq G$ in $K_{\partial U}(\overline{U}, C)$ (that is, there exists a continuous, compact mapping $H : \overline{U} \times [0,1] \to C$ with $H_t(\cdot) : \overline{U} \to C$ belonging to $K_{\partial U}(\overline{U}, C)$ for each $t \in [0,1]$ with $H_0 = F$ and $H_1 = G$). Show that F is essential in $K_{\partial U}(\overline{U}, C)$ if and only if G is essential in $K_{\partial U}(\overline{U}, C)$.

(b) Suppose that $p \in U$. Show that the constant map $F(\overline{U}) = p$ is essential in $K_{\partial U}(\overline{U}, C)$.

(c) Let E be a Fréchet space and C a closed, convex subset of E. Suppose that $F : C \to E$. Show for each $\Omega \subseteq C$ that there exists a closed, convex subset D, depending on F, C and Ω, with $\Omega \subseteq D$ and

$$\overline{\text{co}}(F(D \cap C) \cup \Omega) = D.$$

8.4 Let E be a Fréchet space, C a closed, convex subset of E and K a lattice with a minimal element (denoted by 0). A mapping $\Phi : 2^E \to K$ is called a measure of noncompactness if for any A, B in 2^E

(a) $\Phi(A) = 0$ if and only if A is precompact,
(b) $\Phi(\overline{\text{co}}(A)) = \Phi(A)$,
(c) $\Phi(A \cup B) = \max\{\Phi(A), \Phi(B)\}$.

Let E, C and Φ, a measure of noncompactness of E, be as described above. Suppose that $F : C \to C$ is a continuous, Φ-condensing map (that is, $F : C \to C$ is said to be Φ-condensing provided that if $\Omega \subseteq C$ and $\Phi(F(\Omega)) \geq \Phi(\Omega)$, then Ω is relatively compact). Show that F has a fixed point.

9

Contractive and Nonexpansive Multivalued Maps

In this chapter we present some fixed point results for multivalued contractive and nonexpansive mappings. Throughout this chapter let (X, d) be a metric space and $B(x, r)$ the open ball in X, centred at x with radius r. Also $B(C, r)$ will denote $\bigcup_{x \in C} B(x, r)$, where C is a subset of X. For C and K two nonempty closed subsets of X, we define

$$D(C, K) := \inf\{\epsilon > 0 : C \subseteq B(K, \epsilon),\ K \subseteq B(C, \epsilon)\} \in [0, \infty].$$

D is called the Hausdorff distance.

Let C be a nonempty subset of X. A multivalued mapping $F : C \to X$ with nonempty, bounded, closed values is said to be *contractive* if there exists a constant k, $0 \le k < 1$, with

$$D(F(x), F(y)) \le k\, d(x, y);$$

and is said to be *nonexpansive* if

$$D(F(x), F(y)) \le d(x, y).$$

We begin by presenting the Banach contraction principle for contractive mappings with closed values. This result is due to Nadler. It will follow immediately from the next result.

Theorem 9.1 *Let (X, d) be a complete metric space with $x_0 \in X$ and $r > 0$. Suppose that $F : \overline{B(x_0, r)} \to X$ is a multivalued contraction with nonempty, bounded, closed values such that*

$$\text{dist}(x_0, F(x_0)) < (1 - k)r \tag{9.1}$$

where $0 \le k < 1$ is the constant of contraction. Then F has a fixed point, that is, there exists $x \in \overline{B(x_0, r)}$ with $x \in F(x)$.

Proof We first show (by induction) that there exists a sequence of points $\{x_n\}$ in $\overline{B(x_0, r)}$ with

(a)$_n$ $$x_n \in F(x_{n-1}),$$

and

(b)$_n$ $$d(x_n, x_{n-1}) < k^{n-1}(1-k)r.$$

Now (9.1) guarantees the existence of a point $x_1 \in F(x_0)$ with

$$d(x_1, x_0) < (1-k)r.$$

Next suppose that there exist points x_i satisfying (a)$_i$, (b)$_i$ for $1 \le i \le n$. Notice that

$$D(F(x_n), F(x_{n-1})) \le kd(x_n, x_{n-1}) < k^n(1-k)r,$$

therefore there exists $x_{n+1} \in F(x_n)$ with

$$d(x_{n+1}, x_n) < k^n(1-k)r.$$

By induction there exists a sequence $\{x_n\}$ in $\overline{B(x_0, r)}$ satisfying (a)$_n$ and (b)$_n$. The estimate

$$d(x_{n+p}, x_n) < (1 + k + \cdots + k^{p-1})k^n(1-k)r$$

implies that $\{x_n\}$ is Cauchy and hence converges to a point $x \in \overline{B(x_0, r)}$. On the other hand, since $F(x)$ is closed and

$$D(F(x_n), F(x)) \le kd(x_n, x) \ \left(\text{therefore } \lim_{n\to\infty} \operatorname{dist}(x_{n+1}, F(x)) = 0\right),$$

we have immediately that $x \in F(x)$. □

Our next result is known as Nadler's fixed point theorem.

Theorem 9.2 *Let (X, d) be a complete metric space and $F : X \to X$ a multivalued contraction with nonempty, bounded, closed values. Then F has a fixed point.*

Proof Let $x_0 \in X$. Choose $r > 0$ with

$$\operatorname{dist}(x_0, F(x_0)) < (1-k)r,$$

where $0 \le k < 1$ is the constant of contraction. The result now follows from Theorem 9.1. □

We next show that the property of having a fixed point is invariant by homotopy for contractive multivalued mappings.

Definition 9.1 Let U be an open subset of X, with $F : \overline{U} \to X$ and $G : \overline{U} \to X$ two multivalued contractive maps with nonempty, bounded, closed values; here $\overline{U}$ denotes the closure of U in X. We say that F and G are *homotopic* if there exists $H : \overline{U} \times [0,1] \to X$, a multivalued mapping with nonempty, bounded, closed values, with the following properties:

(a) $H(\cdot, 1) = F$ and $H(\cdot, 0) = G$;
(b) $x \notin H(x,t)$ for every $x \in \partial U$ and $t \in [0,1]$ (here ∂U denotes the boundary of U in X);
(c) there exists α, $0 \le \alpha < 1$, such that $D(H(x,t), H(y,t)) \le \alpha\, d(x,y)$ for all $t \in [0,1]$ and $x, y \in \overline{U}$;
(d) there exists a continuous, increasing function $\phi : [0,1] \to \mathbf{R}$ such that $D(H(x,t), H(x,s)) \le |\phi(t) - \phi(s)|$ for all $t, s \in [0,1]$ and $x, y \in \overline{U}$.

Theorem 9.3 *Let (X, d) be a complete metric space and U an open subset of X. Suppose that $F : \overline{U} \to X$ and $G : \overline{U} \to X$ are two homotopic contractions with nonempty, bounded, closed values and G has a fixed point in U. Then F has a fixed point in U.*

Proof Let H be the homotopy between F and G. Consider the set

$$Q := \{(t,x) \in [0,1] \times U : x \in H(x,t)\}.$$

Notice that Q is nonempty since G has a fixed point. On Q we define the partial order

$$(t,x) \le (s,y) \text{ if and only if } t \le s \text{ and } d(x,y) \le \frac{2(\phi(s) - \phi(t))}{1-\alpha},$$

where α and ϕ are as in Definition 9.1.

Let P be a totally ordered subset of Q. We now show that P has an upper bound in Q. Define

$$t^\star := \sup\{t : (t,x) \in P\}.$$

Take a sequence $\{(t_n, x_n)\}$ in P with

$$(t_n, x_n) \le (t_{n+1}, x_{n+1}) \text{ and } t_n \to t^\star.$$

Notice that

$$d(x_m, x_n) \le \frac{2(\phi(t_m) - \phi(t_n))}{1-\alpha} \text{ for all } m > n.$$

Thus $\{x_n\}$ is a Cauchy sequence and therefore converges to some $x^\star \in \overline{U}$.

It is clear from the continuity of H (see Exercise 9.1) that $(t^\star, x^\star) \in Q$. It is also immediate that

$$(t, x) \leq (t^\star, x^\star) \text{ for every } (t, x) \in P,$$

therefore $(t^\star, x^\star)$ is an upper bound on P. Consequently every totally ordered subset of Q has an upper bound in Q. It follows from Zorn's lemma that Q admits a maximal element $(t_0, x_0) \in Q$.

To complete the proof we need to show that $t_0 = 1$. Suppose that this is not true. Then we can choose $r > 0$ and $t \in (t_0, 1]$ with

$$\overline{B(x_0, r)} \subset U \text{ and } r = \frac{2(\phi(t) - \phi(t_0))}{1 - \alpha}.$$

Next notice that

$$\begin{aligned} \operatorname{dist}(x_0, H(x_0, t)) &\leq \operatorname{dist}(x_0, H(x_0, t_0)) + D(H(x_0, t_0), H(x_0, t)) \\ &\leq \phi(t) - \phi(t_0) < (1 - \alpha)r. \end{aligned}$$

Theorem 9.1 guarantees that $H(\cdot, t)$ has a fixed point $x \in \overline{B(x_0, r)}$. Thus $(t, x) \in Q$ with $(t_0, x_0) < (t, x)$. This however contradicts the maximality of (t_0, x_0). □

We next present a nonlinear alternative of Leray–Schauder type for multivalued contractions in the particular case where X is a Banach space.

Theorem 9.4 *Suppose that U is an open subset of a Banach space X, $0 \in U$ and $F : \overline{U} \to X$ a multivalued contraction with nonempty, closed values such that $F(\overline{U})$ is bounded. Then either*

(A1) *F has a fixed point in $\overline{U}$, or*
(A2) *there exist $x \in \partial U$ and $\lambda \in (0, 1)$ with $x \in \lambda F(x)$*

is true.

Proof Assume that (A2) does not hold and F has no fixed points on ∂U. Then $u \notin \lambda F(u)$ for all $u \in \partial U$ and $\lambda \in [0, 1]$. Let $H : \overline{U} \times [0, 1] \to X$ be given by

$$H(x, t) := tF(x)$$

and let G be the zero map. Apply Theorem 9.3 to deduce that there exists $x \in U$ with $x \in F(x)$. □

As in Chapter 3 we now ask the question whether we can extend Theorem 9.4 to nonexpansive multivalued maps. Let E be a Banach space and K a nonempty, closed subset of E. The following concepts will be needed in the proof of our main result.

Definition 9.2 Let $\{x_n\}$ be a bounded sequence in E. The *asymptotic radius of* $\{x_n\}$ *in* K is the number defined by

$$r(K, \{x_n\}) := \inf_{x \in K} \limsup_{n \to \infty} \|x - x_n\|.$$

Definition 9.3 Let $\{x_n\}$ be a bounded sequence in E. The *asymptotic centre of* $\{x_n\}$ *in* K is the (possibly empty) set defined by

$$A(K, \{x_n\}) := \left\{ x \in K : \limsup_{n \to \infty} \|x - x_n\| \leq r(K, \{x_n\}) \right\}.$$

Definition 9.4 A bounded sequence $\{x_n\}$ is said to be *regular relative to* K if

$$r(K, \{x_n\}) = r(K, \{x_{n_k}\}) \text{ for every subsequence } \{x_{n_k}\} \text{ of } \{x_n\}.$$

Remark 9.1 To become familiar with these concepts we advise the reader at this stage to attempt Exercise 9.2 and Exercise 9.3.

Theorem 9.5 *Let E be a uniformly convex Banach space and U a nonempty, open, bounded, convex subset of E. Suppose that $F : \overline{U} \to E$ is a multivalued nonexpansive mapping with nonempty, compact values. Also assume that there exists a multivalued mapping $H : \overline{U} \times [0,1] \to E$, with nonempty, bounded, closed values, which has the following properties:*

(a) $H(\cdot, 1) = F$;
(b) *$H(\cdot, 0)$ has a fixed point in $\overline{U}$;*
(c) *for every $t \in [0,1)$, there exists α, $0 \leq \alpha < 1$, such that for all x, $y \in \overline{U}$ and all $s \in [0,t]$ we have $D(H(x,s), H(y,s)) \leq \alpha \|x - y\|$;*
(d) *there exists a continuous function $\phi : [0,1] \to \mathbf{R}$ such that for every t, $s \in [0,1]$ we have $D(H(x,t), H(x,s)) \leq |\phi(t) - \phi(s)|$.*

Then either

(A1) *F has a fixed point in $\overline{U}$, or*
(A2) *there exist $t \in [0,1)$ and $x \in \partial U$ with $x \in H(x,t)$*

is true.

Proof Assume that (A2) does not hold (in particular note that this implies that $H(\cdot, 0)$ has no fixed points on ∂U). Now Theorem 9.3 guarantees that for *every* $t \in [0,1)$, $H(\cdot, t)$ has a fixed point in U. Thus we can take sequences $\{t_n\}$ and $\{x_n\}$ such that

$$t_n \in [0,1),\ x_n \in U,\ x_n \in H(x_n, t_n) \text{ and } t_n \to 1.$$

The sequence $\{x_n\}$ is bounded and we can assume that it is regular relative to $\overline{U}$ by Exercise 9.3. Let

$$\{x\} := A(\overline{U}, \{x_n\}) \text{ and } r := r(\overline{U}, \{x_n\}).$$

Since $F(x)$ is compact valued we may choose (see Exercise 9.5) $y_n \in F(x)$ such that

$$\|x_n - y_n\| \le D(H(x_n, t_n), F(x)). \tag{9.2}$$

Again since $F(x)$ is compact valued, some subsequence $\{y_{n_k}\}$ of $\{y_n\}$ converges to $y \in F(x)$. Notice that (9.2), and (c) and (d) in the statement of the theorem, yield

$$\begin{aligned} \|x_{n_k} - y\| &\le \|x_{n_k} - y_{n_k}\| + \|y_{n_k} - y\| \\ &\le D(H(x_{n_k}, t_{n_k}), H(x_{n_k}, 1)) + D(H(x_{n_k}, 1), H(x, 1)) \\ &\quad + \|y_{n_k} - y\| \\ &\le |\phi(t_{n_k}) - \phi(1)| + \|x_{n_k} - x\| + \|y_{n_k} - y\|. \end{aligned}$$

Consequently

$$\limsup_{k\to\infty} \|x_{n_k} - y\| \le \limsup_{k\to\infty} \|x_{n_k} - x\| = r.$$

Now Exercise 9.2 implies that $y = x$ and therefore $x \in F(x)$. □

Theorem 9.6 *Let E be a uniformly convex Banach space, U a nonempty, open, bounded, convex subset of E and $0 \in U$. Suppose that $F : \overline{U} \to E$ is a multivalued nonexpansive mapping with nonempty, compact values such that $F(\overline{U})$ is bounded. Then either*

(A1) *F has a fixed point in $\overline{U}$, or*
(A2) *there exist $x \in \partial U$ and $\lambda \in (0,1)$ with $x \in \lambda F(x)$.*

Proof Let $H : \overline{U} \times [0,1] \to X$ be given by

$$H(x,t) := tF(x)$$

and let G be the zero map. Now apply Theorem 9.5. □

Notes The results in Chapter 9 were taken from Frigon [68].

Exercises

9.1 Let (X, d) be a metric space and C a nonempty subset of X. A multivalued mapping $F : C \to X$ with nonempty values is said to be continuous if it is continuous with respect to the Hausdorff distance D. Show that the mapping H in Definition 9.1 is continuous.

9.2 Let E be a uniformly convex Banach space, K a nonempty, bounded, closed, convex subset of E and $\{x_n\}$ a bounded sequence in E. Show that $A(K, \{x_n\})$ is a singleton.

9.3 Let E be a Banach space. Show that every bounded sequence in E has a subsequence regular relative to K.

9.4 Let (X, d) be a metric space, Ψ_d the collection of nonempty, bounded (that is, d-bounded), closed subsets of X and let D be the Hausdorff distance. Show that the set Ψ_d equipped with D is a metric space.

9.5 Let (X, d) be a metric space, D the Hausdorff metric and A and B nonempty subsets of X.

(a) If B is compact, show that for each $a \in A$, there exists some $b \in B$ with $d(a, b) \leq D(A, B)$.

(b) If A and B are both compact, show that there exist $a \in A$ and $b \in B$ with $d(a, b) = D(A, B)$.

9.6 Show that the fixed point in Theorem 9.2 need not be unique.

9.7 Let E be a Banach space, $F : E \to E$ such that for all $r > 0$, $F|_{\overline{B(0,r)}}$ is a multivalued contraction with nonempty, bounded, closed values. Define

$$\Phi(F) := \{x \in E : x \in \lambda F(x) \text{ for some } \lambda \in [0, 1]\}.$$

Show that either $\Phi(F)$ is unbounded or F has a fixed point.

9.8 Suppose that U is an open subset of a Banach space E, $0 \in U$ and $F : \overline{U} \to X$ is a multivalued contraction with nonempty, closed values such that $F(\overline{U})$ is bounded. Assume that for every $x \in \partial U$, any one of the following conditions is satisfied:

(i) $\|F(x)\| = D(0, F(x)) \leq \|x\|$,

(ii) $\|F(x)\| \leq \operatorname{dist}(x, F(x))$,

(iii) $\|F(x)\| \leq ([\operatorname{dist}(x, F(x))]^2 + \|x\|^2)^{1/2}$,

(iv) $\|F(x)\| \leq \max\{\|x\|, \mathrm{dist}(x, F(x))\}$.

Show that F has a fixed point.

10

Multivalued Maps with Continuous Selections

In this chapter we present some fixed point results for multivalued maps which have continuous selections. To illustrate the methods involved we will, for the convenience of the reader, restrict our attention to *one* particular class of maps, namely the $\Phi^\star$ maps. In addition an application in abstract economies is given.

Suppose that X and Y are Hausdorff topological spaces and $F : X \to Y$. F is said to be *of type* $\Phi^\star$, and we write $F \in \Phi^\star(X, Y)$, if

(1) Y is convex,

(2) $F(x)$ has convex values for all $x \in X$,

(3) there exists a selection $B : X \to Y$ such that $B(x) \neq \varnothing$ for all $x \in X$ and the fibres $B^{-1}(y) = \{z : y \in B(z)\}$ are open (in X) for all $y \in Y$.

Remark 10.1 B is a selection of F if $B(x) \subseteq F(x)$ for all $x \in X$.

Definition 10.1 The topological space X is *paracompact* if X is Hausdorff and every open covering of X has a locally finite open refinement that covers X.

We now begin with a selection theorem for $\Phi^\star$ maps.

Theorem 10.1 *Let X and Y be Hausdorff topological spaces with $F \in \Phi^\star(X, Y)$. If X is paracompact then there exists a continuous, single valued selection $s : X \to Y$ of F.*

Proof Let $B : X \to Y$ be a selection of F. Then

$$\{B^{-1}(y) : y \in Y\}$$

is an open cover of X. Since X is paracompact there exists a locally finite refinement

$$\{U_\alpha\}_{\alpha \in A} \text{ of } \{B^{-1}(y) : y \in Y\}$$

that covers X. For each $\alpha \in A$ fix y_α such that $U_\alpha \subseteq B^{-1}(y_\alpha)$. Let $\{\lambda_\alpha\}_{\alpha \in A}$ be a partition of unity subordinate to $\{U_\alpha\}_{\alpha \in A}$. For each $x \in X$ there exists a finite collection

$$N_x = \{\alpha_1, \ldots, \alpha_n\}$$

with $\lambda_\alpha(x) \neq 0$ if and only if $\alpha \in N_x$. Let $s : X \to Y$ be defined by

$$s(x) := \sum_{\alpha \in N_x} \lambda_\alpha(x) y_\alpha.$$

It is easy to see that $s : X \to Y$ is continuous. In addition $\alpha \in N_x$ implies that $\lambda_\alpha(x) \neq 0$, which in turn implies that $x \in U_\alpha$. Thus

$$\alpha \in N_x \Rightarrow x \in B^{-1}(y_\alpha) \Leftrightarrow y_\alpha \in B(x) \subseteq F(x).$$

Now since $F(x)$ is convex it follows that

$$s(x) \in F(x) \text{ for all } x \in X. \qquad \square$$

We are now in a position to present fixed point theory for $\Phi^\star$ maps.

Theorem 10.2 *Let Q be a convex subset of a Hausdorff locally convex linear topological space E and let $G \in \Phi^\star(Q, Q)$ be a compact map. Then G has a fixed point, that is, there exists $x \in Q$ with $x \in G(x)$.*

Remark 10.2 Recall that $G : Q \to Q$ is compact if $G(Q)$ is relatively compact in Q.

Proof of Theorem 10.2 Choose a compact set X such that

$$G(Q) \subseteq X \subseteq Q.$$

Let $Y := \text{co}(X)$ (that is the convex hull of X). Let $G^\star$ denote the restriction of G to Y. We first show that

$$G^\star \in \Phi^\star(Y, Y).$$

To see this let $B : Q \to Q$ be a selection of G with $B(x) \neq \varnothing$ for all

$x \in Q$ and $B^{-1}(y)$ is open in Q for all $y \in Q$. Let $B^\star$ denote the restriction of B to Y. Notice that for $y \in Y$,

$$\begin{aligned} (B^\star)^{-1}(y) &= \{x \in \text{co}(X) : y \in B^\star(x)\} \\ &= \{x \in \text{co}(X) : y \in B(x)\} \\ &= \text{co}(X) \cap B^{-1}(y) \end{aligned}$$

which is open in $Y = \text{co}(X)$. Consequently $G^\star \in \Phi^\star(Y, Y)$ (note $G^\star(Y) \subseteq X \subseteq Y$). It is also well known (see Exercise 10.1) that $Y = \text{co}(X)$ is paracompact. Now Theorem 10.1 guarantees that $G^\star : Y \to Y$ has a continuous selection $g : Y \to Y$. Thus $g : Y \to Y$ is continuous and compact (since $g(Y) \subseteq G^\star(Y) \subseteq X \subseteq Y$) and therefore the Schauder–Tychonoff theorem (Theorem 8.3) implies that g has a fixed point. Thus G has a fixed point. □

Next we present a nonlinear alternative of Leray–Schauder type for $\Phi^\star$ maps.

Theorem 10.3 *Let Q be a convex subset of a Hausdorff locally convex linear topological space E. Assume that U is relatively open in Q, $u_0 \in U$, $\overline{U}$ (the closure of U in Q) is paracompact and $G \in \Phi^\star(\overline{U}, Q)$ a compact map. Then either*

(A1) *G has a fixed point in $\overline{U}$, or*
(A2) *there exist $u \in \partial U$ and $\lambda \in (0,1)$ with $u \in \lambda G(u) + (1-\lambda)\{u_0\}$*

is true.

Proof Suppose that (A2) does not hold and assume that G does not have a fixed point in ∂U (otherwise we are finished, that is, (A1) occurs). Let $f : \overline{U} \to Q$ (note $\overline{U}$ is paracompact) be a continuous selection of G (guaranteed from Theorem 10.1). Put

$$H := \{x \in \overline{U} : x = \lambda f(x) + (1-\lambda)u_0 \text{ for some } \lambda \in [0,1]\}.$$

Notice that $H \neq \varnothing$ since $u_0 \in U$. In addition H is closed since f is continuous. Moreover H is compact since G is compact map. Since $H \cap \partial U = \varnothing$ there exists a continuous function (note that E is completely regular)

$$\mu : \overline{U} \to [0,1] \text{ with } \mu(H) = 1 \text{ and } \mu(\partial U) = 0.$$

Let

$$J(x) := \begin{cases} \mu(x)f(x) + (1-\mu(x))u_0, & x \in \overline{U}, \\ u_0, & x \in Q \backslash \overline{U}. \end{cases}$$

It is easy to see that $J : Q \to Q$ is continuous. In fact J is compact since G is a compact map. The Schauder–Tychonoff theorem (Theorem 8.3) implies that there exists $w \in Q$ with $w \in J(w)$. In fact $w \in U$ since $u_0 \in U$. Thus

$$w = \mu(w)f(w) + (1 - \mu(w))u_0 = \lambda f(w) + (1 - \lambda)u_0$$

where $0 \leq \lambda = \mu(x) \leq 1$. Consequently $w \in H$ and therefore $\mu(w) = 1$. Thus $w = f(w) \in G(w)$. □

We also have a Furi–Pera type result for $\Phi^\star$ maps.

Theorem 10.4 *Let E be a metrisable (with metric d) locally convex linear topological space with Q a closed, convex subset of E and $0 \in Q$. Assume that $G \in \Phi^\star(Q, E)$ is a compact map and suppose that the following condition is satisfied:*

$$(10.1) \quad \begin{cases} \text{if } \{(x_j, \lambda_j)\}_{j=1}^{\infty} \text{ is a sequence in } \partial Q \times [0,1] \text{ converging to} \\ (x, \lambda) \text{ with } x \in \lambda G(x) \text{ and } 0 \leq \lambda < 1, \text{ then there exists} \\ j_0 \in \{1, 2, \ldots\} \text{ with } \{\lambda_j G(x_j)\} \subseteq Q \text{ for each } j \geq j_0. \end{cases}$$

Then G has a fixed point in Q.

Remark 10.3 Note that Q is paracompact since metrisable spaces are paracompact, and closed subsets of paracompact spaces are paracompact.

Proof of Theorem 10.4 From Theorem 10.1 there exists a continuous selection $f : Q \to E$ of G. Let $r : E \to Q$ be a continuous retraction with $r(z) \in \partial Q$ for $z \in E \backslash Q$ (see Theorem 8.5). Consider

$$X := \{x \in E : x = fr(x)\}.$$

Notice that $X \neq \varnothing$; this follows from the Schauder–Tychonoff theorem (Theorem 8.3) since $fr : E \to E$ is a continuous, compact map (note $fr(E) \subseteq G(Q)$). In addition X is closed and in fact compact since $X \subseteq fr(X) \subseteq G(Q)$.

It remains to show that $X \cap Q \neq \varnothing$. Suppose that $X \cap Q = \varnothing$. Then there exists $\delta > 0$ with $\operatorname{dist}(X, Q) > \delta$. Choose $m \in \{1, 2, \ldots\}$ such that $1 < \delta m$. Fix $i \in \{m, m+1, \ldots\}$. Let

$$U_i := \left\{ x \in E : d(x, Q) < \frac{1}{i} \right\}.$$

Thus $X \cap \overline{U}_i = \emptyset$. Now the nonlinear alternative for compact single valued maps (Theorem 8.4) implies (since $X \cap \overline{U}_i = \emptyset$) that there exists

$$(y_i, \lambda_i) \in \partial U_i \times (0,1) \text{ with } y_i = \lambda_i fr(y_i).$$

Now $\lambda_i fr(y_i) \notin Q$ for $i \in \{m, m+1, \ldots\}$ and therefore

$$\{\lambda_i Gr(y_i)\} \nsubseteq Q \text{ for each } i \in \{m, m+1, \ldots\}. \tag{10.2}$$

Let

$$D := \{x \in E : x = \lambda fr(x) \text{ for some } \lambda \in [0,1]\}.$$

D is compact (since G is compact); therefore we may assume without loss of generality that

$$\lambda_j \to \lambda^\star \text{ and } y_j \to y^\star \in \partial Q.$$

In addition $y^\star = \lambda^\star fr(y^\star)$. Now $\lambda^\star \neq 1$ since $X \cap Q = \emptyset$. Thus $0 \leq \lambda^\star < 1$. But in this case (10.1), with $x_j = r(y_j)$, $x = y^\star = r(y^\star)$, implies that there exists $j_0 \in \{1, 2, \ldots\}$ with $\{\lambda_j Gr(y_j)\} \subseteq Q$ for each $j \geq j_0$. This contradicts (10.2). Thus $X \cap Q \neq \emptyset$, in other words there exists $w \in Q$ with $w \in Gr(w) = G(w)$. □

Theorem 10.2 and Theorem 10.4 will immediately give us coincidence type results for $\Phi^\star$ maps.

Theorem 10.5 *Let E_1 and E_2 be Hausdorff locally convex linear topological spaces, Q a convex subset of E_1 and C a convex subset of E_2. Suppose that $G \in \Phi^\star(Q, C)$ and $F \in \Phi^\star(C, Q)$ are compact maps. Then G and F^{-1} have a coincidence, that is, there exists $(x_0, y_0) \in Q \times C$ with $y_0 \in G(x_0)$ and $x_0 \in F(y_0)$.*

Proof Define the map

$$\Psi(x, y) := F(y) \times G(x) \text{ for } (x, y) \in Q \times C.$$

It is easy to see that $\Psi \in \Phi^\star(Q \times C, Q \times C)$ and Ψ is a compact map since $\Psi(Q \times C) \subseteq F(C) \times G(Q)$. Theorem 10.2 implies that there exists $(x_0, y_0) \in Q \times C$ with $(x_0, y_0) \in \Psi(x_0, y_0)$. □

Theorem 10.6 *Let E be a metrisable locally convex linear topological space with Q and X closed, convex subsets of E and $0 \in Q$. Assume that $G \in \Phi^\star(Q, X)$ and $F \in \Phi^\star(X, E)$ are compact maps and suppose* *that the following condition is satisfied:*

$$(10.3)\quad \begin{cases} \text{if } \{(x_j,\lambda_j)\}_{j=1}^{\infty} \text{ is a sequence in } \partial Q\times[0,1] \text{ converging to} \\ (x,\lambda) \text{ with } x\in\lambda FG(x) \text{ and } 0\le\lambda<1, \text{ then there exists} \\ j_0\in\{1,2,\ldots\} \text{ with } \{\lambda_j FG(x_j)\}\subseteq Q \text{ for each } j\ge j_0. \end{cases}$$

Then G and F^{-1} have a coincidence, that is, there exists $(x_0,y_0)\in Q\times X$ with $y_0\in G(x_0)\cap F^{-1}(x_0)$.

Proof Q is paracompact. Let $g:Q\to X$ be a continuous selection of G (guaranteed from Theorem 10.1) and note that $Fg\in\Phi^\star(Q,E)$ is a compact map. Suppose that $\{(x_j,\lambda_j)\}_{j=1}^{\infty}$ is a sequence in $\partial Q\times[0,1]$ converging to (x,λ) with $x\in\lambda Fg(x)$ and $0\le\lambda<1$. Then $x\in\lambda FG(x)$, therefore (10.3) implies that there exists $j_0\in\{1,2,\ldots\}$ with $\{\lambda_j FG(x_j)\}\subseteq Q$ for each $j\ge j_0$. Thus $\{\lambda_j Fg(x_j)\}\subseteq Q$ for each $j\ge j_0$. Apply Theorem 10.4 to deduce that there exists $x_0\in Q$ with $x_0\in Fg(x_0)$. □

To illustrate the generality of the fixed point theorems derived in this chapter we discuss an application in abstract economies. First however we need to present a generalisation of Theorem 10.2 which is particularly useful in the abstract economy setting.

Theorem 10.7 *Let I be an index set and $\{Q_i\}_{i\in I}$ a family of nonempty, convex sets, each in a locally convex Hausdorff linear topological space E_i. For each $i\in I$, let $G_i\in\Phi^\star(Q,Q_i)$ be a compact map; here $Q:=\prod_{i\in I}Q_i$. Then there exists*

$$x^\star\in G(x^\star)\equiv\prod_{i\in I}G_i(x^\star),$$

that is,

$$x_i^\star\in G_i(x^\star)\ \textit{for all}\ i\in I;$$

here $x_i^\star$ is the projection of $x^\star$ on E_i.

Proof For each $i\in I$, choose a compact set $X_i\subseteq E_i$ with

$$G_i(Q)\subseteq X_i\subseteq Q_i.$$

Let $X:=\prod_{i\in I}X_i$. Note that X is compact. Let $Y:=\text{co}(X)$ and from Exercise 10.1 we see that Y is paracompact. Let $G_i^\star$ denote the restriction of G_i to Y. We now show that

$$G_i^\star \in \Phi^\star(Y, \text{co}(X_i)) \text{ for each } i \in I.$$

(Note that $G_i(Y) \subseteq G_i(Q) \subseteq X_i \subseteq \text{co}(X_i)$.) To see this, for each $i \in I$, let $B_i : Q_i \to Q_i$ be a selection of G_i with $G_i(x) \neq 0$ for all $x \in Q$ and $B^{-1}(y)$ open in Q for all $y \in Q_i$. Let $B_i^\star$ denote the restriction of B to Y. Notice that for each $i \in I$ and $y_i \in \text{co}(X_i)$,

$$\begin{aligned} (B_i^\star)^{-1}(y_i) &= \{x \in \text{co}(X) : y_i \in B_i^\star(x)\} \\ &= \{x \in \text{co}(X) : y_i \in B_i(x)\} \\ &= \text{co}(X) \cap B_i^{-1}(y_i) \end{aligned}$$

which is open in $Y = \text{co}(X)$. Thus $G_i^\star \in \Phi^\star(Y, \text{co}(X_i))$ for each $i \in I$. Theorem 10.1 guarantees that G_i has a continuous selection $g_i : Y \to \text{co}(X_i)$ for each $i \in I$ (in fact $g_i : Y \to X_i$ since $g_i(Y) \subseteq G_i(Y) \subseteq X_i$). Let $g : Y \to Y$ be defined by

$$g(x) := \prod_{i \in I} g_i(x);$$

note that $g(Y) \subseteq \prod_{i \in I} X_i = X \subseteq Y$. Also g is continuous and compact since $g(Y) \subseteq X$. Now the Schauder–Tychonoff theorem (Theorem 8.3) implies that there exists $x^\star \in Y \subseteq Q$ with

$$x^\star = g(x^\star) = \prod_{i \in I} g_i(x^\star) \subseteq \prod_{i \in I} G_i(x^\star). \qquad \square$$

Let I be the set of agents (possibly uncountable). We shall describe an abstract economy by

$$\Gamma = (Q_i, F_i, G_i, P_i)_{i \in I},$$

where for each $i \in I$, Q_i ($\subseteq E_i$) is the choice (or strategy) set, $F_i, G_i : \prod_{i \in I} Q_i \equiv Q \to 2^{Q_i}$ are constraint correspondences and $P_i : Q \to 2^{Q_i}$ is a preference correspondence. A point $x^\star \in Q$ is called an *equilibrium point for* Γ (or a *generalised Nash equilibrium point*) if for each $i \in I$,

$$x_i^\star \in G_i(x^\star) \text{ and } F_i(x^\star) \cap P_i(x^\star) = \varnothing$$

(here $x_i^\star$ is the projection of $x^\star$ on E_i); if such an $x^\star$ exists we say that Γ has an equilibrium point.

Theorem 10.8 *Let $\Gamma = (Q_i, F_i, G_i, P_i)_{i\in I}$ be an abstract economy such that for each $i \in I$ the following conditions hold:*

$$\text{(10.4)} \quad \begin{cases} Q_i \text{ is a nonempty, convex set in a Hausdorff} \\ \text{locally convex linear topological space } E_i; \end{cases}$$

$$\text{(10.5)} \quad \begin{cases} \text{for each } x \in Q \equiv \prod_{i\in I} Q_i, \text{ we have } F_i(x) \neq \varnothing, \\ F_i(x) \subseteq G_i(x) \text{ and } G_i(x) \text{ is convex valued;} \end{cases}$$

$$\text{(10.6)} \quad \begin{cases} \text{for each } y \in Q_i, \text{ the set } [(\mathrm{co}(P_i))^{-1}(y_i) \cup M_i] \cap F_i^{-1}(y_i) \\ \text{is open in } Q; \text{ here } M_i := \{x \in Q : F_i(x) \cap P_i(x)\} = \varnothing; \end{cases}$$

$$\text{(10.7)} \quad G_i : Q \to 2^{Q_i} \text{ is a compact map;}$$

and

$$\text{(10.8)} \quad \begin{cases} \text{for each } x \in Q,\ x_i \notin \mathrm{co}(P_i(x)); \\ \text{here } x_i \text{ is the projection of } x \text{ on } E_i. \end{cases}$$

Then Γ has an equilibrium point $x^\star \in Q$, that is, for each $i \in I$,

$$x_i^\star \in G_i(x^\star) \text{ and } F_i(x^\star) \cap P_i(x^\star) = \varnothing;$$

here $x_i^\star$ is the projection of $x^\star$ on E_i.

Proof For each $i \in I$ let

$$N_i := \{x \in Q : F_i(x) \cap P_i(x) \neq \varnothing\}$$

and for each $x \in Q$ let

$$I(x) := \{i \in I : F_i(x) \cap P_i(x) \neq \varnothing\}.$$

For each $i \in I$, define the correspondences $A_i, B_i : Q \to 2^{Q_i}$ by

$$A_i(x) := \begin{cases} \mathrm{co}(P_i(x)) \cap F_i(x) \text{ if } i \in I(x) \text{ (that is, if } x \in N_i), \\ F_i(x) \text{ if } i \notin I(x) \end{cases}$$

and

$$B_i(x) := \begin{cases} \mathrm{co}(P_i(x)) \cap G_i(x) \text{ if } i \in I(x), \\ G_i(x) \text{ if } i \notin I(x). \end{cases}$$

It is easy to see (use (10.5)) that for each $i \in I$ and $x \in Q$

$$B_i(x) \text{ is convex valued, } A_i(x) \neq 0 \text{ and } A_i(x) \subseteq B_i(x).$$

In addition for each $i \in I$ and $y_i \in Q_i$ we have

$$\begin{aligned}(A_i)^{-1}(y_i) &= \{[(\mathrm{co}(P_i))^{-1}(y_i) \cap F_i^{-1}(y_i)] \cap N_i\} \cup [F_i^{-1}(y_i) \cap M_i] \\ &= [(\mathrm{co}(P_i))^{-1}(y_i) \cap F_i^{-1}(y_i)] \cup [F_i^{-1}(y_i) \cap M_i] \\ &= [(\mathrm{co}(P_i))^{-1}(y_i) \cup M_i] \cap F_i^{-1}(y_i)\end{aligned}$$

which is open in Q by (10.6). Thus $B_i \in \Phi^\star(Q, Q_i)$ for each $i \in I$. In addition (10.7) implies that B_i is a compact map for each $i \in I$. Now Theorem 10.7 implies that there exists $x^\star \in Q$ with $x_i^\star \in B_i(x^\star)$ for all $i \in I$ (here $x_i^\star$ is the projection of $x^\star$ on E_i). Note that if $i \in I(x^\star)$ for some $i \in I$, then

$$F_i(x^\star) \cap P_i(x^\star) \neq \varnothing \text{ and } x_i^\star \in \mathrm{co}(P_i(x^\star)) \cap G_i(x^\star)$$

(in particular $x_i^\star \in \mathrm{co}(P_i(x^\star))$). This contradicts (10.8). Thus $i \notin I(x^\star)$ for all $i \in I$. Consequently we have that $F_i(x^\star) \cap P_i(x^\star) = \varnothing$ and $x_i^\star \in G_i(x^\star)$ for all $i \in I$. □

Notes The results in Chapter 10 were adapted from Ben-El-Mechaiekh, Deguire and Granas [18], and O'Regan [142, 143]. Other maps of this type are discussed in [52, 87, 126, 178].

Exercises

10.1 Let D be a nonempty, compact subset of a Hausdorff topological space E. Show that $\mathrm{co}(D)$ is σ-compact (hence Lindelöf and paracompact).

10.2 Let X and Y be Hausdorff topological spaces with $F \in \Phi^\star(X, Y)$. If X is compact show that there exist a finite collection $\{y_1, \ldots, y_n\}$ of elements of Y and a continuous single valued selection $s : X \to Y$ of F with $s(X) \subseteq \mathrm{co}\{y_1, \ldots, y_n\}$.

10.3 Let X be a nonempty, paracompact Hausdorff topological space and Y a nonempty, convex subset of a topological vector space. Suppose that $S, T : X \to 2^Y$ are correspondences such that

(a) for each $x \in X$, $\mathrm{co}(S(x)) \subseteq T(x)$ and $S(x) \neq \varnothing$,
(b) for each $y \in Y$, $S^{-1}(y)$ is open in X.

Show that T has a continuous selection.

10.4 Let $\{X_i\}_{i \in I}$ be a family of nonempty, convex sets, each in a locally convex Hausdorff topological vector space E_i, where I is an index set. For each $i \in I$, let D_i be a nonempty compact subset of X_i and $S_i, T_i : X = \prod_{i \in I} X_i \to 2^{D_i}$ be such that for each $i \in I$,

(a) for each $x \in X$, $\mathrm{co}(S_i(x)) \subseteq T_i(x)$ and $S_i(x) \neq \varnothing$,
(b) for each $y_i \in D_i$, $S_i^{-1}(y_i)$ is open in X.

Show that there exists
$x^\star \in D = \prod_{i \in I} D_i$ with $x^\star \in T(x^\star) = \prod_{i \in I} T_i(x^\star)$.

11

Multivalued Maps with Closed Graph

In this chapter we extend the Schauder–Tychonoff theorem to the multivalued case. The basic results in this chapter are due to S. Kakutani, I. L. Glicksberg and K. Fan.

We begin this chapter by presenting Sperner's combinatorial lemma. A set $\{x_0, \ldots, x_n\}$ in a Euclidean space is said to be *affinely independent* if

$$\sum_{i=0}^{n} \lambda_i x_i = 0 \text{ and } \sum_{i=0}^{n} \lambda_i = 0 \text{ imply } \lambda_0 = \cdots = \lambda_n = 0.$$

Definition 11.1 Let $v_0, v_1, \ldots, v_n$ be an affinely independent set of $n+1$ points in a Euclidean space E. Their convex hull

$$\left\{ x \in E : x = \sum_{i=0}^{n} \lambda_i v_i,\ 0 \leq \lambda_i \leq 1,\ \sum_{i=0}^{n} \lambda_i = 1 \right\}$$

is called the *(closed) n-simplex* and is denoted by $v_0 v_1 \ldots v_n$. The points $v_0, v_1, \ldots, v_n$ are called the *vertices* of the simplex.

For $0 \leq k \leq n$ and $0 \leq i_0 < i_1 < \cdots < i_k \leq n$, the k-simplex $v_{i_0} v_{i_1} \ldots v_{i_k}$ is a subset of the n-simplex $v_0 v_1 \ldots v_n$; it is called a k dimensional *face* (or simply *k-face*) of $v_0 v_1 \ldots v_n$. In addition if $y = \sum_{i=0}^{n} \lambda_i v_i$ we let $\chi(y) = \{i : \lambda_i > 0\}$.

Suppose that $T = \{\tau_0, \ldots, \tau_m\}$ is a collection of n-simplexes in $v_0 v_1 \ldots v_n$ such that

(a) $\bigcup_{i=1}^{m} \tau_i = v_0 v_1 \ldots v_n$
and

(b) $\tau_i \cap \tau_j$ is either empty or a common face of τ_i and τ_j $(i, j = 1, \ldots, m)$.

Then T is called a *triangulation* (or *simplicial subdivision*) of $v_0v_1\dots v_n$. The mesh of the subdivision is the diameter of the largest subsimplex.

Let T be a triangulation. A vertex of a simplex in T is called a vertex of T. An $(n-1)$-face of a simplex from T is called a *boundary* $(n-1)$-*simplex of* T if it is a face of exactly one n-simplex of T.

We now sketch the proof of Sperner's combinatorial lemma (the proof here is based on an argument by Ky Fan).

Theorem 11.1 *Let T be a triangulation of an n-simplex $v_0v_1\dots v_n$. Let ϕ be a function which assigns to each vertex v of T one of the integers $0,1,\dots,n$ (that is, $\phi(v)\in\{0,1,\dots,n\}$) such that whenever*

$$(11.1)\qquad v\in v_{i_0}v_{i_1}\dots v_{i_k}\ \text{then}\ \phi(v)\in\{i_0,i_1,\dots,i_k\}$$

for $0\le k\le n$ and $0\le i_0<i_1<\dots<i_k\le n$ (that is, such that $\phi(v)\in\chi(v)$ – we say that ϕ is a proper labelling of the triangulation). Then the number of n-simplexes in T whose vertices receive $n+1$ different values is odd.

Proof We may think of T as a house; the n-simplexes are the rooms of the house. A room will be considered a *good* room if its $n+1$ vertices are labelled with $n+1$ different integers, otherwise it will be called a *bad* room. If the vertices of an $(n-1)$-face of a simplex τ of T are labelled with the integers $0,1,\dots,n-1$, then that face is considered to be a door to the room τ. A door which is a boundary $(n-1)$-simplex of T is called an external door. It is clear that a good room has exactly one door and a bad room has either no doors or two doors. We will now trace paths through the house going through each door exactly once.

To *start* a path either we enter a room through an external door or we leave a good room through its unique door. Every time we enter a room through a door and see a second door in the room, we go out through the second door. This will lead us to either the outside of the house or a new room. According to this rule, our path will stop either when we get to the outside of the house through an external door or when we get to a good room. After going through each door exactly once by tracing various paths, we pair off the two ends of each path. In this way each good room is paired with either another good room or an external door. In addition each external door is paired with either another external door or a good room. This pairing implies that the number of good rooms and the number of external doors are congruent modulo 2 (that is, these numbers have the same parity). To complete the proof (Exercise 11.9) we note (use induction) that the number of boundary $(n-1)$-simplexes in T with vertices labelled $0, 1, \dots, n-1$ is odd. □

We now use Sperner's lemma to present the theorem of Knaster, Kuratowski and Mazurkiewicz (KKM theorem).

Theorem 11.2 *Let $\{F_0, \dots, F_n\}$ be a family of $n+1$ closed subsets of an n-simplex $v_0 v_1 \dots v_n$. Suppose that for each $0 \le k \le n$ and $0 \le i_0 < i_1 < \cdots < i_k \le n$ we have*

$$v_{i_0} v_{i_1} \dots v_{i_k} \subseteq F_{i_0} \cup F_{i_1} \cup \cdots \cup F_{i_k}. \tag{11.2}$$

Then $\bigcap_{i=0}^{n} F_i \ne \varnothing$.

Proof Let T be an arbitrary triangulation of the simplex $v_0 v_1 \dots v_n$. For each vertex v of T we assign the integer $\phi(v) \in \{0, 1, \dots, n\}$ with

$$v \in F_{\phi(v)} \text{ and } \phi(v) \in \{i_0, \dots, i_k\} \text{ whenever } v \in \{v_{i_0}, v_{i_1}, \dots, v_{i_k}\};$$

notice that (11.2) guarantees that this possible. Now Theorem 11.1 implies that there is a simplex $w_0 w_1 \dots w_n$ of T with

$$\phi(w_i) = i \text{ for } i = 0, 1, \dots, n.$$

Then

$$w_i \in F_{\phi(w_i)} = F_i \text{ for } i = 0, 1, \dots, n.$$

Consider the sequence of triangulations $\{T^{(m)}\}$ of $v_0 v_1 \dots v_n$ such that the mesh of $T^{(m)}$ tends to zero as $m \to \infty$. For each m, there exists an n-simplex $w_0^{(m)} w_1^{(m)} \dots w_n^{(m)}$ in T with $w_i^{(m)} \in F_i$ for $i = 0, 1, \dots, n$. The compactness of $v_0 v_1 \dots v_n$ guarantees that we can choose a subsequence $\{m_j\}$ of $\{m\}$ with $m_1 < m_2 < \cdots$ and such that

$$\lim_{j\to\infty} w_i^{(m_j)} = q_j \text{ exists for } i = 0, 1, \dots, n.$$

Now since the mesh of $T^{(m_j)}$ tends to zero as $j \to \infty$, it follows immediately that $q_0 = q_1 = \cdots = q_n$. Thus

$$\lim_{j\to\infty} w_i^{(m_j)} = q \text{ for all } i = 0, 1, \dots, n.$$

Since F_i is closed for each $i = 0, 1, \dots, n$ we have that $q \in F_i$ for $i = 0, 1, \dots, n$. Consequently $\bigcap_{i=0}^{n} F_i \ne \varnothing$. □

We next present an infinite dimensional version of the KKM theorem. This will be used to establish Ky Fan's minimax theorem which is the crucial result that is needed to establish the analogue of the Schauder–Tychonoff theorem for multivalued maps.

Definition 11.2 Let Q be a convex set in a vector space X, A a nonempty subset of Q and $F : A \to 2^Q$ (here 2^Q denotes the set of all subsets of Q). The family $\{F(x) : x \in A\}$ is said to be a *KKM covering for* Q if

$$\mathrm{co}\{x : x \in N\} \subseteq \bigcup_{x\in N} F(x)$$

for any finite set $N \subseteq A$.

Theorem 11.3 *Let Q be a convex set in a Hausdorff topological vector space X, A a nonempty subset of Q, $F : A \to 2^Q$ and $\{F(x) : x \in A\}$ a KKM covering for Q. If there exists an $a \in A$ with $\overline{F(a)}$ compact, then*

$$\bigcap_{x\in A} \overline{F(x)} \neq \varnothing.$$

Proof From a standard result in topology it suffices (since $\overline{F(a)}$ is compact) to show that the family $\{\overline{F(x)} : x \in A\}$ has the finite intersection property. Let $\{x_0, \ldots, x_n\} \subseteq A$ and let $\sigma = v_0 v_1 \ldots v_n$ be an n-simplex. Define $\psi : \sigma \to Q$ by

$$\psi\left(\sum_{i=0}^{n} \mu_i v_i\right) = \sum_{i=0}^{n} \mu_i x_i \tag{11.3}$$

for each $\mu_i \geq 0$ $(i = 0, \ldots, n)$ with $\sum_{i=0}^{n} \mu_i = 1$. Now ψ is continuous, therefore the set

$$G_i = \psi^{-1}(\overline{F(x_i)}) \text{ is closed in } \sigma$$

for each $i = 0, \ldots, n$. If $0 \leq k \leq n$ and $0 \leq i_0 < i_1 < \cdots < i_k \leq n$, we *claim* that

$$v_{i_0} v_{i_1} \ldots v_{i_k} \subseteq \bigcup_{j=0}^{k} G_{i_j}.$$

If the claim is true, then Theorem 11.2 guarantees that $\bigcap_{i=0}^{n} G_i \neq \varnothing$.

However,

$$\bigcap_{j=0}^{n} G_j = \bigcap_{i=0}^{n} \psi^{-1}(\overline{F(x_i)}) = \psi^{-1}\left(\bigcap_{i=0}^{n} \overline{F(x_i)}\right),$$

and therefore $\bigcap_{i=0}^{n} \overline{F(x_i)} \neq \varnothing$ and we are finished.

It remains to prove the claim. Since $\{F(x) : x \in A\}$ is a KKM covering for Q we have

$$\operatorname{co}\{x_{i_0}, x_{i_1}, \ldots, x_{i_k}\} \subseteq \bigcup_{j=0}^{k} F(x_{i_j}) \subseteq \bigcup_{j=0}^{k} \overline{F(x_{i_j})}.$$

This together with (11.3) gives

$$v_{i_0} v_{i_1} \ldots v_{i_k} = \operatorname{co}\{v_{i_0}, v_{i_1}, \ldots, v_{i_k}\} = \psi^{-1}(\operatorname{co}\{x_{i_0}, x_{i_1}, \ldots, x_{i_k}\})$$

and therefore

$$v_{i_0} v_{i_1} \ldots v_{i_k} \subseteq \psi^{-1}\left(\bigcup_{j=0}^{k} \overline{F(x_{i_j})}\right) = \bigcup_{j=0}^{k} \psi^{-1}\left(\overline{F(x_{i_j})}\right) = \bigcup_{j=0}^{k} G_{i_j}. \quad \square$$

We are now in a position to prove Ky Fan's minimax theorem. A real valued function ϕ on a topological space X is *lower* (respectively *upper*) *semicontinuous* if the set $\{x \in X : \phi(x) \leq \lambda\}$ (respectively $\{x \in X : \phi(x) \geq \lambda\}$) is closed in X for each $\lambda \in \mathbf{R}$. If Q is a convex set in a vector space then a real valued function ϕ on Q is said to be *quasiconcave* (respectively *quasiconvex*) if $\{x \in Q : \phi(x) > \lambda\}$ (respectively $\{x \in Q : \phi(x) < \lambda\}$) is convex for each $\lambda \in \mathbf{R}$.

Theorem 11.4 *Let K be a nonempty, convex, compact subset of a Hausdorff topological vector space X and ϕ a real valued function on the product space $K \times K$ satisfying the following conditions:*

(11.4) *for each fixed $x \in K$, $\phi(x, \cdot)$ is lower semicontinuous on K*

and

(11.5) *for each fixed $y \in K$, $\phi(\cdot, y)$ is quasiconcave on K.*

Then there exists $y^\star \in K$ with

$$\phi(x, y^\star) \leq \sup_{z \in K} \phi(z, z) \text{ for all } x \in K$$

$$\left(\text{and therefore } \min_{y \in K} \sup_{x \in K} \phi(x, y) \leq \sup_{x \in K} \phi(x, x)\right).$$

Proof Let $\lambda := \sup_{x \in K} \phi(x, x)$. We may assume that $\lambda \neq \infty$. For each $x \in K$ let

$$F(x) := \{y \in K : \phi(x, y) \leq \lambda\}.$$

Condition (11.4) guarantees that each $F(x)$ is closed and hence compact in K (note that K is compact). We *claim* that

$$\{F(x) : x \in K\} \text{ is a KKM covering for } K.$$

If the claim is true then Theorem 11.3 guarantees that $\bigcap_{x \in K} F(x) \neq \varnothing$.

Take $y^\star \in \bigcap_{x \in K} F(x)$ and we are finished.

It remains to prove the claim. Suppose it is not true. Then there exist $\{x_1, \ldots, x_n\} \subseteq K$ and $\alpha_i \geq 0$ $(i = 0, 1, \ldots, n)$ with $\sum_{i=0}^{n} \alpha_i = 1$ such that

$$w = \sum_{i=0}^{n} \alpha_i x_i \notin \bigcup_{i=0}^{n} F(x_i).$$

This together with the definition of $F(x)$ yields

$$\phi\left(x_i, \sum_{i=0}^{n} \alpha_i x_i\right) = \phi(x_i, w) > \lambda \text{ for } i = 0, 1, \ldots, n. \tag{11.6}$$

Finally (11.6) together with the quasiconcavity of $\phi(\cdot, w)$ guarantees that $\phi(w, w) > \lambda$ – a contradiction. □

Definition 11.3 Let X and Y be topological spaces and $F : X \to 2^Y$ be a set valued map (that is, for each $x \in X$, $f(x) \in 2^Y$). We say that F is *upper semicontinuous* at $x_0 \in X$ if for each open set G with $f(x_0) \subseteq G$, there exists an open neighbourhood $N(x_0)$ of x_0 with $F(x) \subseteq G$ for all $x \in N(x_0)$. F is upper semicontinuous on X if it is upper semicontinuous at every point of X.

Theorem 11.5 *Let X be a Hausdorff topological vector space, Q a nonempty subset of X and $F : Q \to 2^Q$ upper semicontinuous with $F(x)$ nonempty and bounded for each $x \in Q$. Then for any $g \in X'$ (dual) the map $\phi_g : Q \to \mathbf{R}$, defined by*

$$\phi_g(x) := \sup_{x \in F(y)} Re\,\langle x, y\rangle \quad (\textit{note } \langle x, g\rangle = g(x)),$$

is upper seimcontinuous in the sense of real valued functions.

Proof Fix $y_0 \in Q$. Let $\epsilon > 0$ be given and let

$$U_\epsilon := \left\{ x \in Q : |\langle x, g \rangle| < \frac{\epsilon}{2} \right\}.$$

Notice that U_ϵ is an open neighbourhood of 0. Since $F(y_0) + U_\epsilon$ is an open set containing $F(y_0)$, it follows from the upper semicontinuity of F at y_0 that there exists a neighbourhood $N(y_0)$ of y_0 in Q with

$$F(y) \subseteq F(y_0) + U_\epsilon \text{ for all } y \in N(y_0).$$

Thus for each $y \in N(y_0)$ we have that

$$\begin{aligned}
\phi_g(y) &= \sup_{x \in F(y)} Re\,\langle x, g \rangle \leq \sup_{x \in F(y_0)+U_\epsilon} Re\,\langle x, g \rangle \\
&\leq \sup_{x \in F(y_0)} Re\,\langle x, g \rangle + \sup_{x \in U_\epsilon} Re\,\langle x, g \rangle \\
&< \phi_g(y_0) + \epsilon,
\end{aligned}$$

therefore ϕ_g is upper semicontinuous. □

We are now in a position to prove the analogue of the Schauder–Tychonoff theorem in the multivalued case, namely Ky Fan's fixed point theorem. In the proof we will use the following separation theorem (the proof of which follows from the Hahn–Banach theorem and can be found in any standard functional analysis text).

Theorem 11.6 *Suppose that A and B are disjoint, nonempty, convex sets in a Hausdorff locally convex linear topological space X. If A is compact and B is closed, then there exist $f \in X'$ and $\gamma \in \mathbf{R}$ with*

$$\max f(A) < \gamma \leq \inf f(B).$$

Theorem 11.7 *Let K be a nonempty, convex, compact subset of a Hausdorff locally convex linear topological space X and let $F : K \to 2^K$ be upper semicontinuous with $F(x)$ a nonempty, closed, convex subset of K for each $x \in K$. Then there exists a $y^\star \in K$ with $y^\star \in F(y^\star)$.*

Proof Suppose that the result is not true, that is, suppose $y \notin F(y)$ for each $y \in K$. Now for each $y \in K$, Theorem 11.5 guarantees that there exists $f_y \in X'$ with

$$Re\,\langle y, f_y \rangle - \sup_{x \in F(y)} Re\,\langle x, f_y \rangle > 0. \tag{11.7}$$

For each $g \in X'$, let

$$V(g) := \left\{ y \in K : Re\langle y, g\rangle - \sup_{x \in F(y)} \langle x, g\rangle > 0 \right\}.$$

Notice that (11.7) ensures that $K = \bigcup_{g \in X'} V(g)$. In addition Theorem 11.5 implies that $V(g)$ is open in K. The compactness of K guarantees the existence of

$$g_1, g_2, \ldots, g_n \in X' \text{ with } K = \bigcup_{i=1}^{n} V(g_i).$$

Let $\{\lambda_1, \ldots, \lambda_n\}$ be a partition of unity on K subordinate to $\{V(g_1), \ldots, V(g_n)\}$ (let $V_i = V(g_i)$ for $i = 1, \ldots, n$), that is, $\lambda_1, \ldots, \lambda_n$ are continuous, nonnegative, real valued functions on K with λ_i vanishing on $K \backslash V_i$ for each $i = 1, \ldots, n$ and $\sum_{i=1}^{n} \lambda_i(x) = 1$ for all $x \in K$. Let $\phi : K \times K \to \mathbf{R}$ be given by

$$\phi(x, y) := \sum_{i=1}^{n} \lambda_i(y) Re\langle y - x, g_i\rangle.$$

It is easy to see that all the conditions of Ky Fan's minimax theorem (Theorem 11.4) are satisfied and therefore there exists $y_0 \in K$ with

$$\phi(x, y_0) \le 0 \text{ for all } x \in K,$$

that is,

$$\sum_{i=1}^{n} \lambda_i(y_0) Re\langle y_0 - x, g_i\rangle \le 0 \text{ for all } x \in K. \tag{11.8}$$

Suppose that $i \in \{1, 2, \ldots, n\}$ is such that $\lambda_i(y_0) > 0$. Then $y_0 \in V(g_i)$ (since λ_i vanishes on $K \backslash V_i$) and consequently

$$Re\langle y_0, g_i\rangle > \sup_{x \in F(y_0)} \langle x, g_i\rangle \ge Re\langle x_0, g_i\rangle$$

for all $x_0 \in F(y_0)$ (that is, $Re\langle y_0 - x_0, g_i\rangle > 0$ for all $x_0 \in F(y_0)$). Thus

$$\lambda_i(y_0) Re\langle y_0 - x_0, g_i\rangle > 0 \text{ whenever } \lambda_i(y_0) > 0 \text{ (for } i = 1, \ldots, n)$$

for all $x_0 \in F(y_0)$. Since $\lambda_i(y_0) > 0$ for at least one $i \in \{1, 2, \ldots\}$, it follows that

$$\sum_{i=1}^{n} \lambda_i(y_0) Re\langle y_0 - x_0, g_i\rangle > 0 \text{ for all } x_0 \in F(y_0).$$

This contradicts (11.8). □

We now present some results where we assume compactness on the operator instead of compactness on the space K.

Theorem 11.8 *Let K be a nonempty, closed, convex subset of a complete Hausdorff locally convex linear topological space and let $F : K \to 2^K$ be an upper semicontinuous, compact map (that is, $F(K)$ is a relatively compact subset of K) with $F(x)$ a nonempty, closed, convex subset of K for each $x \in K$. Then there exists a $y^\star \in K$ with $y^\star \in F(y^\star)$.*

Proof Let $Q := \overline{\text{co}}(\overline{F(K)})$. It is well known (Exercise 11.1) that Q is a nonempty, convex, compact subset of K. In addition $F(Q) \subseteq \overline{F(K)} \subseteq Q$. Theorem 11.7 guarantees that there exists $y^\star \in Q$ with $y^\star \in F(y^\star)$. □

Theorem 11.9 *Let K be a nonempty, closed, convex subset of a complete Hausdorff locally convex linear topological space and let $F : K \to 2^K$ be a closed (that is, having closed graph), compact map with $F(x)$ a nonempty, closed, convex subset of K for each $x \in K$. Then F has a fixed point in K.*

Proof Let $Q := \overline{\text{co}}(\overline{F(K)})$. Notice that $F : Q \to 2^Q$ with Q compact. Also, $F|_Q$ has closed graph, therefore Exercise 11.5 ensures that $F|_Q$ is upper semicontinuous. Now apply Theorem 11.7. □

Finally in this chapter we establish a nonlinear alternative of Leray–Schauder type for closed maps.

Theorem 11.10 *Let E be a complete Hausdorff locally convex linear topological space, U an open subset of E and $0 \in U$. Suppose that $F : \overline{U} \to 2^E$ is a closed, compact map with $F(x)$ a nonempty, closed, convex subset of E for each $x \in \overline{U}$. Then either*

(A1) *F has a fixed point in $\overline{U}$, or*
(A2) *there are a $u \in \partial U$ and a $\lambda \in (0, 1)$ with $u \in \lambda F(u)$.*

Proof Suppose that (A2) does not hold and F has no fixed points on ∂U (otherwise we are finished). Consider

$$A := \{x \in \overline{U} : x \in \lambda F(x) \text{ for some } \lambda \in [0, 1]\}.$$

Notice that $A \neq \varnothing$ is closed. To see this let (x_α) be a net in A (that is, $x_\alpha \in \lambda_\alpha F(x_\alpha)$ for some $\lambda_\alpha \in [0,1]$) with $x_\alpha \to x_0 \in \overline{U}$. Without loss of generality, assume that $\lambda_\alpha \to \lambda_0 \in (0,1]$. Since $x_\alpha \in A$ there exists $y_\alpha \in F(x_\alpha)$ with $x_\alpha = \lambda_\alpha y_\alpha$. Now $x_\alpha \to x_0$ and $y_\alpha \to \dfrac{1}{\lambda_0} x_0$. The closedness of F implies that $\dfrac{1}{\lambda_0} x_0 \in F(x_0)$ and therefore $x_0 \in A$. Thus A is closed. In addition $F : \overline{U} \to 2^E$ being a compact map implies that A $(= \overline{A})$ is compact.

Notice that $A \cap \partial U = \emptyset$. Since E is completely regular, A is compact and ∂U is closed, there exists a continuous function $\mu : \overline{U} \to [0,1]$ with $\mu(A) = 1$ and $\mu(\partial U) = 0$. Let

$$N(x) := \begin{cases} \mu(x) F(x), & x \in \overline{U}, \\ \{0\}, & x \in E \backslash \overline{U}. \end{cases}$$

It is immediate that $N : E \to 2^E$ is a closed, compact map (see Exercise 11.1) and $N(x)$ has nonempty, closed, convex values for each $x \in E$. Theorem 11.9 ensures the existence of an $x \in E$ with $x \in N(x)$. Notice that $x \in U$ since $0 \in U$. Thus

$$x \in \mu(x) F(x) = \lambda F(x) \text{ where } 0 \leq \lambda = \mu(x) \leq 1.$$

Consequently $x \in A$, therefore $\mu(x) = 1$ and we are finished. □

Remark 11.1 We remark that the results in Chapter 6 and Chapter 7 extend to upper semicontinuous, compact maps. We leave the details to the reader.

Notes The approach in Chapter 11 is adapted from Fan [64] and Wong [188]. We refer the reader to Aliprantis and Border [3], and Zeidler [191] for other approaches. Theorem 11.10 is taken from O'Regan [141].

Exercises

11.1 Let E be a complete Hausdorff locally convex linear topological space. Show that the closed convex hull of a compact subset of E is compact.

11.2 Show that an upper semicontinuous map $F : X \to 2^Y$ is closed if either

(i) F is closed valued and Y is regular, or

(ii) F is compact valued and Y is Hausdorff

hold.

11.3 Let A be a subset of a regular space X and let $F : A \to 2^X$ be an upper semicontinuous, closed valued map. Show that the set of fixed points of F is closed.

11.4 Let Q be a closed, convex subset of a Fréchet space E, $0 \in Q$, and let $F : Q \to 2^E$ be a closed, compact map with $F(x)$ a nonempty, closed, convex subset of E for each $x \in Q$. In addition assume that

$$\begin{cases} \text{if } \{(x_j, \lambda_j)\}_{j=1}^{\infty} \text{ is a sequence in } \partial Q \times [0,1] \text{ converging to } (x, \lambda) \\ \text{with } x \in \lambda F(x) \text{ and } 0 \leq \lambda < 1, \text{ then there exists } j_0 \in \{1, 2, \ldots\} \\ \text{with } \{\lambda_j F(x_j)\} \subseteq Q \text{ for each } j \geq j_0 \end{cases}$$

is satisfied. Show that F has a fixed point in Q.

11.5 Show that a closed valued map with compact Hausdorff range space (codomain) is closed if and only if it is upper semicontinuous.

11.6 Let K be a nonempty, convex, compact subset of a Haudorff topological vector space X and let ψ be a real valued continuous function defined on $K \times K$. If for each fixed $y \in K$, the function $\psi(\cdot, y)$ is quasiconvex on K, show that there exists $y^\star \in K$ with

$$\psi(x, y^\star) \geq \psi(y^\star, y^\star) \text{ for all } x \in K.$$

11.7 Complete the details in Remark 11.1.

11.8 Let A be a nonempty, convex subset of a complete Hausdorff locally convex linear topological space and let $F : A \to 2^K$ be an upper semicontinuous map with $F(x)$ a nonempty, closed, convex subset of A for each $x \in K$. In addition suppose that $F(A)$ is contained in some compact subset C of A. Show that F has a fixed point. (Note that we do not assume that A is closed!)

11.9 Complete the proof in Theorem 11.1 (that is, supply the induction argument).

11.10 Let X be a Fréchet space with metric d. Suppose that $K \subseteq X$ is compact, V is a neighbourhood of 0 and $\epsilon > 0$. Show that there exists a convex neighbourhood of 0, W, such that $W \subseteq V$ and also such that whenever $x \in K$ and $y \in W$ then $d(x + y, x) < \epsilon$.

11.11 Let X be a Fréchet space, $C \subseteq X$ closed and convex, and let $F : C \to 2^X$ be an upper seimcontinuous map with $F(x)$ a nonempty, compact, convex subset of X for each $x \in C$. Then given $\epsilon > 0$ and d a metric on X, show that there exists a continuous, single valued mapping $f : C \to X$ such that if $x \in C$ then there are a $y \in C$ and $z \in F(y)$ such that $d(x, y) < \epsilon$ and $d(f(x), z) < \epsilon$.

12
Degree Theory

In our final chapter we introduce the concept of the degree of a map. It will be used here to provide another approach to presenting (i) fixed point theory, and (ii) continuation principles.

We begin by defining the degree of a map defined on subsets of $\mathbf{R}^n$. Here

$$\mathbf{R}^n := \{x = (x_1, \ldots, x_n) : x_i \in \mathbf{R} \text{ for } i = 1, \ldots, n\}$$

with $|x| := \max\{|x_i| : i = 1, \ldots, n\}$; ρ denotes the induced distance (induced by $|\cdot|$) and

$$B(y, r) := \{x : \rho(x, y) < r\}$$

is the open ball with centre y and radius r. Let D be a bounded open subset of $\mathbf{R}^n$ and $p \in \mathbf{R}^n$. $C(\overline{D})$ is the linear subspace of continuous functions from $\overline{D}$ into $\mathbf{R}^n$ with norm

$$\|f\| := \sup_{x \in D} |f(x)|;$$

$f'(x)$ is the derivative of the function f at the point x; the symbol '$_{,j}$' denotes the operator $\dfrac{\partial}{\partial x_j}$ and $J_f(x)$ is the Jacobian of f at x, that is,

$$J_f(x) := \det f'(x) = \det(f_{i,j}(x)).$$

We say that $f \in C^1(\overline{D})$ if $f \in C(\overline{D})$ and there is an extension $\tilde{f}$ of f defined on an open set $\mathcal{D}(f)$ containing $\overline{D}$ such that f has continuous first order partial derivatives in $\mathcal{D}(f)$; the norm on $C^1(\overline{D})$ is

$$\|f\|_1 := \sup_{\substack{x \in D \\ 1 \le i \le n}} |f_i(x)| + \sup_{\substack{x \in D \\ 1 \le i,j \le n}} |f_{i,j}(x)|.$$

$C^2(\overline{D})$ is the subspace of $C^1(\overline{D})$ consisting of those functions for which the corresponding extensions $\tilde{f}$ have continuous partial derivatives of the second order in $\mathcal{D}(f)$. Finally $C_0^r(\overline{D})$ $(r = 1, 2)$ is the subspace of $C^r(\overline{D})$ consisting of those functions whose support is contained in D (the support of f, denoted by $\operatorname{supp}(f)$, is $\overline{\{x : f(x) \neq 0\}}$).

Definition 12.1 A point $x \in \mathbf{R}^n$ for which $f(x) = p$ is called a *p*-point of f. For $f \in C(\overline{D})$, $f^{-1}(p)$ is then the collection of p-points of f in $\overline{D}$.

Definition 12.2 Let $\phi \in C^1(\overline{D})$. We say that x is a *critical point* of ϕ if $J_\phi(x) = 0$; then $\phi(x)$ is a *critical value* of ϕ. The set of critical points of ϕ in $\overline{D}$ is denoted by $Z_\phi(\overline{D})$ or Z_ϕ. The set of critical values, $\phi(Z_\phi)$, is called the crease of ϕ.

Theorem 12.1 *If $\phi \in C^1(\overline{D})$ and $p \notin \operatorname{crease} \phi$, then $\phi^{-1}(p)$ is finite.*

Proof We *claim* that the p-points of ϕ in $\overline{D}$ are isolated. Suppose the claim is true. Notice that $\phi^{-1}(p)$ is closed and also compact. Since the p-points of ϕ in $\overline{D}$ are isolated, there exists an open covering $\{A_\alpha\}_{\alpha \in \mathcal{A}}$ of $\phi^{-1}(p)$ with

$$\phi^{-1}(p) \subseteq \bigcup_{\alpha \in \mathcal{A}} A_\alpha$$

and with each A_α containing at most one p-point of ϕ in $\overline{D}$. We can now extract a finite subcovering and as a result, $\phi^{-1}(p)$ is finite.

It remains to prove the claim. If this were not true then we would have a sequence (x_n) in $\overline{D}$ convergent to x_0 say, with $\phi(x_n) = p$ for all n. Then $x_0 \in \overline{D}$, $\phi(x_0) = p$ and

$$(12.1)\quad 0 = \phi(x_n) - \phi(x_0) = \phi'(x_0)(x_n - x_0) + o(|x_n - x_0|) \text{ as } n \to \infty.$$

Since $\phi'(x_0)$ is nonsingular, there is an $r > 0$ with

$$(12.2)\qquad |\phi'(x_0)(u)| \geq r|u| \text{ for } u \in \mathbf{R}^n.$$

For n sufficiently large we have from (12.1) that

$$|\phi'(x_0)(x_n - x_0)| < \frac{1}{2} r |x_n - x_0|,$$

and this contradicts (12.2). □

Definition 12.3 Suppose that $\phi \in C^1(\overline{D})$, $p \notin \phi(\partial D)$ and $p \notin \operatorname{crease} \phi$.

Define the *degree of ϕ at p relative to D* to be $d(\phi, D, p)$, where

$$d(\phi, D, p) := \sum_{x \in \phi^{-1}(p)} \operatorname{sign} J_\phi(x).$$

Remark 12.1

(i) ∂D denotes the boundary of D.

(ii) Since $p \notin \operatorname{crease} \phi$, Theorem 12.1 guarantees that the summation in the definition is finite.

We have immediately the following result.

Theorem 12.2 *Let I denote the identity map. If $p \in D$ then $d(I, D, p) = 1$, whereas if $p \notin \overline{D}$ then $d(I, D, p) = 0$.*

Our goal is to remove the restrictions that $\phi \in C^1(\overline{D})$ and $p \notin \operatorname{crease} \phi$ in Definition 12.3. We first give some definitions (the justifications will follow).

Definition 12.4 If $\phi \in C^1(\overline{D})$ and $p \notin \phi(\partial D)$ but $p \in \operatorname{crease} \phi$, define $d(\phi, D, p)$ to be $d(\phi, D, q)$, where q is any point such that

$$q \notin \operatorname{crease} \phi \text{ and } |q - p| < \rho(p, \phi(\partial D)).$$

Remark 12.2 Since ∂D is compact, $\phi(\partial D)$ is closed and therefore as a result $\rho(p, \phi(\partial D)) > 0$ since $p \notin \phi(\partial D)$.

To justify Definition 12.4 we need to show that we can find a q as described. In addition we need to prove that the degree does not depend on the particular point q chosen.

Definition 12.5 Suppose that $\phi \in C(\overline{D})$ and $p \notin \phi(\partial D)$. Define $d(\phi, D, p)$ to be $d(\psi, D, p)$ where ψ is any function in $C^1(\overline{D})$ satisfying

$$|\phi(x) - \psi(x)| < \rho(p, \phi(\partial D)) \text{ for } x \in \overline{D}.$$

To justify Definition 12.5 we need to show that the degree does not depend on the function ψ chosen.

We now gather together the results that will be needed to justify the above definitions.

Theorem 12.3 *Suppose that $\phi \in C^1(\overline{D})$ and $p \notin \phi(Z_\phi) \cup \phi(\partial D)$. Then there is a $\delta > 0$, depending on p and ϕ, such that if $\|\psi - \phi\|_1 < \delta$ then*

$$p \notin \psi(Z_\psi) \cup \psi(\partial D) \text{ and } d(\psi, D, p) = d(\phi, D, p).$$

Proof If $\phi^{-1}(p) = \varnothing$ take $\delta = \frac{1}{2}\rho(p, \phi(\partial D))$. Then if $\|\psi - \phi\|_1 < \delta$ we have $|\psi(x) - p| > \delta$ for $x \in \overline{D}$, that is, ψ has no p-points in $\overline{D}$. Consequently $d(\phi, D, p) = d(\psi, D, p)$ from Definition 12.3.

Suppose next that $\phi^{-1}(p) = \{a_1, \ldots, a_k\}$ (we know that $\phi^{-1}(p)$ is finite since p is not a critical value of ϕ). We can take (Exercise 12.1) disjoint balls $B_i := B(a_i, r)$ and make a sequence of choices of r and δ such that eventually every $\psi \in C^1(\overline{D})$ satisfying $\|\psi - \phi\|_1 < \delta$ has exactly one p-point in each B_i and none elsewhere, and J_ϕ and J_ψ have the same sign in each B_i.

With r chosen as above, there is an $\epsilon > 0$ with $|\phi(x) - p| \geq \epsilon$ for $x \in A \equiv \overline{D} \backslash B(r)$ where $B(r) := B(a_1, r) \cup \cdots \cup B(a_k, r)$. By decreasing δ further if necessary we can ensure that $|\phi(x) - \psi(x)| \leq \frac{\epsilon}{2}$ for $x \in \overline{D}$. Thus if $x \in A$,

$$|\psi(x) - p| \geq |\phi(x) - p| - |\phi(x) - \psi(x)| \geq \frac{\epsilon}{2}$$

and therefore $\psi^{-1}(p) \subseteq B(r)$. Since no point of $B(r)$ is a critical point of ψ, we also have that $p \notin \operatorname{crease} \psi$. Consequently

$$d(\psi, D, p) = \sum_{x \in \psi^{-1}(p)} \operatorname{sign} J_\psi(x) = \sum_{i=1}^{k} \operatorname{sign} J_\phi(a_i) = d(\phi, D, p). \qquad \square$$

Next we state, for the convenience of the reader, a version of Sard's well known theorem on manifolds which will be particularly useful here.

Theorem 12.4 *Let $\phi \in C^1(\overline{D})$. Then the crease $\phi(Z_\phi)$ has zero measure in $\mathbf{R}^n$.*

Remark 12.3 Since $\phi(Z_\phi)$ has measure zero, it cannot have interior points.

If X is a subset of $\mathbf{R}^n$, we can write $C^1(X, \mathbf{R}^m)$ for the vector space of continuously differentiable functions from X into $\mathbf{R}^m$, and $C_0^1(X, \mathbf{R}^m)$ for the subset consisting of those functions whose support is compact. We now state two well known results. The proofs are not difficult (Exercise 12.2) and can be found in any standard analysis book.

Theorem 12.5 *Let $\phi \in C^2(\overline{D})$ and $v \in C_0^1(\mathbf{R}^n, \mathbf{R}^n)$ with* supp(v) *disjoint from $\phi(\partial D)$. Then there is a function $u \in C_0^1(\overline{D})$ with*

$$(\operatorname{div} u)(x) = J_\phi(x) \cdot (\operatorname{div} v)(\phi(x)).$$

Theorem 12.6 *Let $f \in C_0^1(\mathbf{R}^n, \mathbf{R})$ and $K = \operatorname{supp}(f)$. Let $\gamma(s)$ be a path in $\mathbf{R}^n$ such that the tube*

$$A \equiv \{k + \gamma(s) : k \in K \text{ and } 0 \leq s \leq 1\}$$

is contained in D. Then there is a $v \in C_0^1(\overline{D})$ with

$$(\operatorname{div} v)(x) = f(x - \gamma(0)) - f(x - \gamma(1)).$$

Our next result expresses the degree of a map as an integral.

Theorem 12.7 *Suppose that $\phi \in C^1(\overline{D})$, $p \notin \phi(\partial D)$ and $p \notin \operatorname{crease} \phi$. Let $f_\epsilon : \mathbf{R}^n \to \mathbf{R}$ be a continuous function with*

$$K_\epsilon \equiv \operatorname{supp}(f_\epsilon) \subseteq B(0, \epsilon) \text{ and } \int_{\mathbf{R}^n} f_\epsilon(x)\, dx = 1.$$

Then there is an ϵ_0, depending on p and ϕ, such that if $0 < \epsilon < \epsilon_0$ then

$$d(\phi, D, p) = \int_D f_\epsilon(\phi(x) - p) J_\phi(x)\, dx. \tag{12.3}$$

Proof Suppose that $\phi^{-1}(p) = \{a_1, \ldots, a_k\}$ (note that $p \notin \operatorname{crease} \phi$). For sufficiently small ϵ there are disjoint neighbourhoods $A_i(\epsilon)$ of a_i in D such that for each i,

$$\phi(A_i(\epsilon)) = B(p, \epsilon) \text{ and } \phi|_{A_i} \text{ is one-to-one.}$$

In addition choose ϵ such that the $A_i(\epsilon)$ are disjoint from ∂D and $J_\phi \neq 0$ in them. As a result

$$\operatorname{supp}(f_\epsilon(\phi(\cdot) - p)) \subseteq \bigcup_{i=1}^k A_i(\epsilon),$$

and therefore since J_ϕ is one sign in each $A_i(\epsilon)$ we have

$$\begin{aligned} \int_D f_\epsilon(\phi(x) - p) J_\phi(x)\, dx &= \sum_{i=1}^k \int_{A_i} f_\epsilon(\phi(x) - p) J_\phi(x)\, dx \\ &= \sum_{i=1}^k \operatorname{sign} J_\phi(a_i) \int_{K_\epsilon} f_\epsilon(y)\, dy \\ &= \sum_{i=1}^k \operatorname{sign} J_\phi(a_i) = d(\phi, D, p). \end{aligned}$$ □

We are now in a position to prove the last theorem that will be needed in the justification of Definition 12.4.

Theorem 12.8 *Let $\phi \in C^1(\overline{D})$ and suppose that p_1 and p_2 are in the same component of $\mathbf{R}^n \backslash \phi(\partial D)$ with neither in the crease of ϕ. Then*

$$d(\phi, D, p_1) = d(\phi, D, p_2).$$

Proof Suppose to begin with that $\phi \in C^2(\overline{D})$. Since $\phi(\partial D)$ is closed, we have that $\mathbf{R}^n \backslash \phi(\partial D)$ is an open subset of $\mathbf{R}^n$ and therefore from a standard topology course we know that its connected components are path connected. Let Ω be the component of $\mathbf{R}^n \backslash \phi(\partial D)$ containing p_1 and p_2. There exists a path $\gamma(s)$ in Ω with $\gamma(0) = p_1$ and $\gamma(1) = p_2$. Since $\{\gamma(s) : 0 \le s \le 1\}$ is compact there exists an $\epsilon_1 > 0$ such that the ϵ_1-neighbourhood of γ is contained in Ω. Let $\epsilon < \epsilon_1$ and let f_ϵ be chosen such that $d(\phi, D, p_1)$ and $d(\phi, D, p_2)$ are given by (12.3). Next let

$$K_{s,\epsilon} := \{z + \gamma(s) : z \in \operatorname{supp}(f_\epsilon)\}.$$

Notice that $K_{s,\epsilon} \subseteq \Omega$ for $0 \le s \le 1$. Theorem 12.6 guarantees a $v \in C_0^1(\overline{D})$ with $\operatorname{supp}(v) \subseteq \Omega$ and

$$(\operatorname{div} v)(x) = f_\epsilon(x - p_1) - f_\epsilon(x - p_2). \tag{12.4}$$

Theorem 12.5 (note that $\phi \in C^2(\overline{D})$ and $\operatorname{supp}(v)$ is disjoint from $\phi(\partial D)$) guarantees a $u \in C_0^1(\overline{D})$ with

$$(\operatorname{div} u)(x) = J_\phi(x) \cdot (\operatorname{div} v)(\phi(x)). \tag{12.5}$$

Now using (12.4), (12.5) and the divergence theorem we obtain

$$\begin{aligned} d(\phi, D, p_1) &= \int_D f_\epsilon(\phi(x) - p_1) J_\phi(x)\, dx \\ &= \int_D f_\epsilon(\phi(x) - p_2) J_\phi(x)\, dx + \int_D J_\phi(x) \cdot (\operatorname{div} v)(\phi(x))\, dx \\ &= \int_D f_\epsilon(\phi(x) - p_2) J_\phi(x)\, dx + \int_D (\operatorname{div} u)(x)\, dx \\ &= d(\phi, D, p_2). \end{aligned}$$

It remains to consider the case when ϕ is no longer in $C^2(\overline{D})$. Of course there is a sequence $\{\phi_n\}_{n=1}^\infty$ with

$$\phi_n \in C^2(\overline{D}) \text{ and } \phi_n \to \phi \text{ in } C^1(\overline{D}) \text{ as } n \to \infty.$$

Let $\gamma(s)$ be as above and let $\delta = \rho(\gamma, \phi(\partial D))$ (note that $\delta > 0$ since γ and $\phi(\partial D)$ are compact). If $\|\phi - \phi_n\| < \dfrac{\delta}{2}$, then for $x \in \partial D$ and

$0 \leq s \leq 1$ we have for this n,

$$|\phi_n(x) - \gamma(s)| \geq |\phi(x) - \gamma(s)| - |\phi(x) - \phi_n(x)| > \frac{\delta}{2};$$

thus p_1 and p_2 are in the same component of $\mathbf{R}^n \backslash \phi_n(\partial D)$. Choosing n sufficiently large we obtain from Theorem 12.3 (applied twice) that

$$d(\phi, D, p_1) = d(\phi_n, D, p_1) = d(\phi_n, D, p_2) = d(\phi, D, p_2). \qquad \square$$

Justification of Definition 12.4 From Theorem 12.4 every ball $B(p, r)$ contains points $q \notin \operatorname{crease} \phi$. The component of p in $\mathbf{R}^n \backslash \phi(\partial D)$ contains $B(p, \rho(p, \phi(\partial D)))$. Consequently Theorem 12.8 guarantees that $d(\phi, D, q)$ has the same value for all $q \notin \operatorname{crease} \phi$ satisfying $|q - p| < \rho(p, \phi(\partial D))$.

We will need one more definition and theorem before we justify Definition 12.5.

Definition 12.6 A C^1 *homotopy* between ϕ and ψ of $C^1(\overline{D})$ is a function $H : \overline{D} \times [0, 1] \to \mathbf{R}^n$ such that $H_0 = \phi$, $H_1 = \psi$, $H_t \in C^1(\overline{D})$ for $0 \leq t \leq 1$ and $\|H_t - H_s\|_1 \to 0$ as $s \to t$; here H_t denotes the function $x \mapsto H(x, t)$.

Theorem 12.9 *Let $\phi \in C^1(\overline{D})$.*

(i) *$d(\phi, D, \cdot)$ is constant on components of $\mathbf{R}^n \backslash \phi(\partial D)$.*
(ii) *If $p \notin \phi(\partial D)$ there is an $\epsilon > 0$, depending on p and ϕ, with $d(\phi, D, p) = d(\psi, D, p)$ when $\|\phi - \psi\|_1 < \epsilon$.*
(iii) *Let $H(x, t)$ be a C^1 homotopy between ϕ and ψ with $p \notin H(\partial D, t)$ for all $t \in [0, 1]$. Then $d(\phi, D, p) = d(\psi, D, p)$.*

Proof

(i) Let Ω be a component of $\mathbf{R}^n \backslash \phi(\partial D)$ with p_1, $p_2 \in \Omega$. For $i = 1, 2$ choose (Theorem 12.4) $q_i \notin \operatorname{crease} \phi$ with $|q_i - p_i| < \rho(p_i, \phi(\partial D))$. Note that $q_1, q_2 \in \Omega$ and Definition 12.4 (applied twice) with Theorem 12.8 give

$$d(\phi, D, p_1) = d(\phi, D, q_1) = d(\phi, D, q_2) = d(\phi, D, p_2).$$

(ii) Let $\delta = \rho(p, \phi(\partial D))$ and choose q such that $|q - p| < \delta/2$ and $q \notin \operatorname{crease} \phi$ (Theorem 12.4). Also choose $\epsilon < \delta/2$ such that $\|\phi - \psi\|_1 < \epsilon$

implies $q \notin \text{crease}\,\psi$ and $d(\phi, D, q) = d(\psi, D, q)$ (Theorem 12.3). Note that if $x \in \partial D$ then

$$|p - \psi(x)| \geq |p - \phi(x)| - |\phi(x) - \psi(x)| > \frac{\delta}{2},$$

therefore p and q are in the same component of $\mathbf{R}^n \backslash \psi(\partial D)$ (also p and q are in the same component of $\mathbf{R}^n \backslash \phi(\partial D)$). Applying part (i) twice we obtain

$$d(\phi, D, p) = d(\phi, D, q) = d(\psi, D, q) = d(\psi, D, p).$$

(iii) $d(H_\epsilon, D, p)$ is defined for all $t \in [0, 1]$ and it is easy to see from part (ii) that it is a continuous function from $[0, 1]$ into $\mathbf{R}$. This together with the fact that d is integer valued guarantees the result. □

Justification of Definition 12.5 Let $\delta = \rho(p, \phi(\partial D))$ and suppose for $i = 1, 2$ that $\psi_i \in C^1(\overline{D})$ with $\|\phi - \psi_i\| < \delta$. Define the C^1 homotopy

$$h_t(x) := t\psi_1(x) + (1 - t)\psi_2(x) \text{ for } x \in \overline{D} \text{ and } 0 \leq t \leq 1.$$

Notice that

$$\begin{aligned} |h_t(x) - \phi(x)| &= |t(\psi_1(x) - \phi(x)) + (1 - t)(\psi_2(x) - \phi(x))| \\ &< t\delta + (1 - t)\delta = \delta, \end{aligned}$$

therefore if $x \in \partial D$ then

$$|p - h_t(x)| \geq |p - \phi(x)| - |\phi(x) - h_t(x)| > \delta - \delta = 0.$$

Thus $p \notin h_t(\partial D)$ for $0 \leq t \leq 1$ and therefore $d(\psi_1, D, p) = d(\psi_2, D, p)$ from Theorem 12.9 (iii). Consequently $d(\psi, D, p)$ is the same for all $\psi \in C^1(\overline{D})$ if $|\phi(x) - \psi(x)| < \delta$ for $x \in \overline{D}$.

Theorem 12.10 *Let $\phi \in C(\overline{D})$ and $p \notin \phi(\partial D)$. If $d(\phi, D, p)$ is nonzero then there is a $q \in D$ with $\phi(q) = p$.*

Proof If $p \notin \phi(\overline{D})$ take $\psi \in C^1(\overline{D})$ with $\|\phi - \psi\| < \rho(p, \phi(\overline{D}))$. For $x \in \overline{D}$ note that

$$|p - \psi(x)| \geq |p - \phi(x)| - |\phi(x) - \psi(x)| > 0.$$

Thus $p \notin \psi(\overline{D})$ and therefore $d(\psi, D, p) = 0$ from Definition 12.3. As a result $d(\phi, D, p) = 0$ from Definition 12.5 – a contradiction. Hence $p \in \phi(D)$, that is, there exists $q \in D$ with $\phi(q) = p$. □

Let X and Y be topological spaces. Two continuous functions f and g are said to be *homotopic* (written $f \simeq g$) if there is a continuous function $h : [0,1] \times X \to Y$ with $h(0,x) = f(x)$ and $h(1,x) = g(x)$ for all $x \in X$.

Theorem 12.11

(i) *Suppose that* $\phi \in C(\overline{D})$ *and* $p \notin \phi(\partial D)$. *If* $\|\psi - \phi\| < \rho(p, \phi(\partial D))$ *then* $d(\psi, D, p)$ *is defined and equals* $d(\phi, D, p)$.

(ii) *If* $h(t,x) := h_t(x)$ *is a homotopy and* $p \notin h_t(\partial D)$ *for* $0 \le t \le 1$, *then* $d(h_t, D, p)$ *is independent of* $t \in [0,1]$.

Proof

(i) If $x \in \partial D$ then

$$|p - \psi(x)| \ge |p - \phi(x)| - |\phi(x) - \psi(x)| > 0.$$

Thus $p \notin \psi(\partial D)$ and therefore $d(\psi, D, p)$ is defined. Let $\sigma \in C^1(\overline{D})$ be such that

$$\|\sigma - \psi\| + \|\psi - \phi\| < \rho(p, \phi(\partial D)).$$

This implies that

$$\|\sigma - \phi\| \le \|\sigma - \psi\| + \|\psi - \phi\| < \rho(p, \phi(\partial D)),$$

therefore Definition 12.5 implies that $d(\phi, D, p) = d(\sigma, D, p)$. In addition since

$$\rho(p, \phi(\partial D)) \le \rho(p, \psi(\partial D)) + \|\phi - \psi\|$$

we have $\|\sigma - \psi\| < \rho(p, \psi(\partial D))$. Once again Definition 12.5 implies that $d(\psi, D, p) = d(\sigma, D, p)$. As a result,

$$d(\psi, D, p) = d(\phi, D, p) \text{ if } \|\psi - \phi\| < \rho(p, \phi(\partial D)).$$

(ii) Notice that $d(h_t, D, p)$ is defined for all $t \in [0,1]$. It is easy to see from part (i) that $t \mapsto d(h_t, D, p)$ is a continuous function from $[0,1]$ into $\mathbf{R}$. This together with the fact that d is integer valued implies that $d(h_t, D, p)$ is a constant. □

We now use degree theory to give an alternative proof of Brouwer's fixed point theorem (Theorem 4.3).

Theorem 12.12 *Let* B^n *denote the closed unit ball in* $\mathbf{R}^n$. *Then every continuous map* $\psi : B^n \to B^n$ *has a fixed point.*

Proof If $\psi(x_0) = x_0$ for some $x_0 \in \partial B^n$ then we are finished.

For the remainder of the proof we suppose that $\psi(x) \neq x$ for $x \in \partial(\operatorname{int} B^n) = \partial B^n$. Consider

$$h_t(x) = x - t\psi(x) \text{ for } x \in B^n \text{ and } 0 \le t \le 1.$$

If $x \in \partial(\operatorname{int} B^n)$ and $0 \le t < 1$, then $t\psi(x) \in \operatorname{int}(B^n)$ and therefore

$$h_t(x) \neq 0 \text{ for } x \in \partial(\operatorname{int} B^n) \text{ and } 0 \le t < 1.$$

Also $0 \notin h_1(\partial(\operatorname{int} B^n))$ since $x \neq \psi(x)$ for $x \in \partial(\operatorname{int} B^n)$. We can apply Theorem 12.11 (ii) to deduce that

$$d(I - \psi, \operatorname{int}(B^n), 0) = d(I, \operatorname{int}(B^n), 0); \tag{12.6}$$

here I denotes the identity map. Notice that Theorem 12.2 guarantees that $d(I, \operatorname{int}(B^n), 0) = 1$ and this together with (12.6) yields

$$d(I - \psi, \operatorname{int}(B^n), 0) = 1.$$

Theorem 12.10 guarantees that there is an $x \in \operatorname{int}(B^n)$ with $(I-\psi)(x) = 0$, that is, $x = \psi(x)$. □

The following nonlinear alternative of Leray–Schauder type is also easily deduced from degree theory.

Theorem 12.13 *Let D be a bounded, open subset of $\mathbf{R}^n$ and $\phi \in C(\overline{D})$. Suppose that $p \in D$ is such that*

$$\phi(x) - p \neq \mu(x - p) \text{ for all } x \in \partial D \text{ and } \mu > 1. \tag{12.7}$$

Then ϕ has a fixed point in $\overline{D}$.

Proof Assume that ϕ has no fixed points in ∂D (otherwise we are finished). Consider

$$h_t(x) = x - p - t(\phi(x) - p) \text{ for } x \in \overline{D} \text{ and } 0 \le t \le 1.$$

If $h_t(x) = 0$ for $x \in \partial D$ and $0 < t < 1$ then

$$\phi(x) - p = \mu(x - p) \text{ with } \mu = \frac{1}{t};$$

which contradicts (12.7). Note also that $h_1(x) \neq 0$ for $x \in \partial D$ and $h_0(x) \neq 0$ for $x \in \partial D$ since $p \in D$. Theorem 12.11 (ii) with $p \in D$ guarantees that

$$d(I - \phi, D, 0) = d(I - p, D, 0) = 1;$$

here I is the identity map. Now apply Theorem 12.10. □

Next we define the degree of a map defined on a normed linear space $(X, \|\cdot\|)$. Here D will denote an open, bounded subset of X and p will be a point of X. As before, ρ will denote the induced distance (induced by $\|\cdot\|$). We shall consider maps $\phi = I - F$ where I is the identity and $F : \overline{D} \to X$ is continuous and compact. Suppose also that $p \notin \phi(\partial D)$. In our definition of degree we will use as before $\rho(p, \phi(\partial D))$. We first show

$$(12.8) \qquad \delta = \rho(p, \phi(\partial D)) > 0.$$

If $\delta := \inf\{\|p - \phi(x)\| : x \in \partial D\} = 0$ then there exists a sequence (x_n) in ∂D with $\phi(x_n) \to p$ as $n \to \infty$. Since $F : \overline{D} \to X$ is a compact map we deduce that the set $\{F(x_n)\}_{n=1}^{\infty}$ is relatively compact and therefore has a convergent subsequence. That is, there exists a subsequence S of $\mathbf{N}$ with $F(x_n) \to y$ (say) as $n \to \infty$ in S. Notice that $y \in \overline{F(\overline{D})}$ and

$$x_n = F(x_n) + \phi(x_n) \to y + p \text{ as } n \to \infty \text{ in } S.$$

Note that $y + p \in \partial D$ since $x_n \in \partial D$ and ∂D is closed. The continuity of F now implies $y = F(y + p)$ and therefore $\phi(y + p) = p$. That is, $p \in \phi(\partial D)$ – a contradiction. Thus (12.8) is true.

Next take ϵ with $0 < \epsilon < \delta = \rho(p, \phi(\partial D))$. The Schauder approximation theorem (Theorem 4.12) guarantees that there is a mapping $F_\epsilon : \overline{D} \to X$ with finite dimensional range such that

$$\|F(x) - F_\epsilon(x)\| < \epsilon \text{ for } x \in \overline{D}.$$

Let H_ϵ be the finite dimensional normed space spanned by $F_\epsilon(\overline{D})$ and p, that is, $H_\epsilon = \operatorname{span}\{F_\epsilon(\overline{D}), p\}$. In addition let

$$D_\epsilon = D \cap H_\epsilon \text{ and } \phi_\epsilon(x) = x - F_\epsilon(x) \text{ for } x \in \overline{D}.$$

Clearly D_ϵ is a bounded, open subset of H_ϵ and $\partial_\epsilon D_\epsilon \subseteq \partial D$ (here $\partial_\epsilon D_\epsilon$ is the boundary of D_ϵ in H_ϵ). In addition $\phi_\epsilon(\overline{D}_\epsilon) \subseteq H_\epsilon$ and for $x \in \partial D$

$$\|x - F_\epsilon(x) - p\| \geq \|x - F(x) - p\| - \|F(x) - F_\epsilon(x)\| > \delta - \epsilon > 0;$$

thus $d(\phi_\epsilon, D_\epsilon, p)$ is defined.

We now *claim*

$$(12.9) \qquad d(\phi_\epsilon, D_\epsilon, p) \text{ is independent of } \epsilon.$$

To see this take ϵ and η both in $(0, \delta)$. Let H_ϵ, H_η be as defined above and let

$$H_\mu := \operatorname{span}\{H_\epsilon,\ H_\eta\} \text{ and } D_\mu := D \cap H_\mu.$$

Exercise 12.5 now guarantees that

$$d(\phi_\epsilon, D_\epsilon, p) = d(\phi_\epsilon, D_\mu, p) \tag{12.10}$$

and

$$d(\phi_\eta, D_\eta, p) = d(\phi_\eta, D_\mu, p). \tag{12.11}$$

Consider

$$h_t(x) = t\phi_\epsilon(x) + (1-t)\phi_\eta(x) \text{ for } x \in \overline{D}_\mu \text{ and } 0 \le t \le 1.$$

Notice that

$$\begin{aligned} \|h_t(x) - \phi(x)\| &\le t\|\phi_\epsilon(x) - \phi(x)\| + (1-t)\|\phi_\eta(x) - \phi(x)\| \\ &< t\epsilon + (1-t)\eta < \delta \end{aligned}$$

and therefore for $x \in \partial D_\mu$ we have

$$\|h_t(x) - p\| \ge \|\phi(x) - p\| - \|\phi(x) - h_t(x)\| > 0.$$

Thus

$$d(\phi_\epsilon, D_\mu, p) = d(\phi_\eta, D_\mu, p). \tag{12.12}$$

Now (12.10), (12.11) and (12.12) guarantee that (12.9) is true.

Finally if V is any finite dimensional space containing H_ϵ where $0 < \epsilon < \delta$, then with $D_V = D \cap V$, Exercise 12.5 guarantees that

$$d(\phi_\epsilon, D_V, p) = d(\phi_\epsilon, D_\epsilon, p).$$

We are now in a position to give the definition of the degree (Leray–Schauder) of maps of the form $I - F$ where I is the identity and F is continuous and compact.

Definition 12.7 Let D be an open, bounded subset of a normed linear space $X = (X, \|\cdot\|)$ and $\phi = I - F$ where $F : \overline{D} \to X$ is continuous and compact. Also let $p \in X \backslash \phi(\partial D)$. Take $\hat{\phi} = I - \hat{F}$ where $\hat{F}$ is a continuous mapping defined in $\overline{D}$ with finite dimensional range chosen such that

$$\|F(x) - \hat{F}(x)\| < \rho(p, \phi(\partial D)) \text{ for } x \in \overline{D}.$$

Choose a finite dimensional linear space V to contain $\hat{F}(\overline{D})$ and p and let $D_V = D \cap V$. Define

$$d(\phi, D, p) = d(\hat{\phi}, D_V, p).$$

Theorem 12.14 *Let D be an open, bounded subset of a normed linear space X. If $p \in D$ then $d(I, D, p) = 1$ whereas if $p \notin \overline{D}$ then $d(I, D, p) = 0$.*

Proof In Definition 12.7 take $\hat{F} = 0$ and the space V to contain p. Now apply Theorem 12.2. □

Theorem 12.15 *Let D be an open, bounded subset of a normed linear space $X = (X, \|\cdot\|)$ and $\phi = I - F$ where $F : \overline{D} \to X$ is continuous and compact. Suppose that $p \notin \phi(\partial D)$ and $d(\phi, D, p) \neq 0$. Then there is an $x \in D$ with $\phi(x) = p$.*

Proof For all sufficiently small ϵ, we can choose as in the Schauder approximation theorem (Theorem 4.12) a mapping F_ϵ with finite dimensional range such that $\|F(x) - F_\epsilon(x)\| < \epsilon$ for $x \in \overline{D}$. Definition 12.7 together with $d(\phi, D, p) \neq 0$ implies for n sufficiently large (that is for $n \geq n_0$, $n_0 \in \mathbf{N}$, say) that there exists an $x_n \in D$ with

$$x_n - F_{\frac{1}{n}}(x_n) = p.$$

Now $\{F(x_n)\}_{n=n_0}^{\infty}$ is relatively compact, therefore there is a subsequence S of $\{n_0, n_0 + 1, \ldots\}$ with $F(x_n) \to y$, say, as $n \to \infty$ in S. In addition,

$$\|x_n - F(x_n) - p\| = \left\| F_{\frac{1}{n}}(x_n) - F(x_n) \right\| < \frac{1}{n},$$

therefore $x_n \to y + p$ as $n \to \infty$ in S. From the continuity of F we have $y = F(y + p)$, therefore $\phi(y + p) = p$. □

Definition 12.8 Let D be an open, bounded subset of a normed linear space $X = (X, \|\cdot\|)$ and suppose h maps $[0, 1]$ into $K(\overline{D})$; here $K(\overline{D})$ denotes the set of continuous, compact maps from $\overline{D}$ into X. We say that h is a *homotopy of compact transformations* on $\overline{D}$ if given $\epsilon > 0$ there is a $\delta > 0$, depending on ϵ, with

$$\|(h(t))(x) - (h(s))(x)\| < \epsilon \text{ for } x \in \overline{D} \text{ and } |t - s| < \delta.$$

Theorem 12.16 (Invariance under homotopy) *Let D be an open, bounded subset of a normed linear space $X = (X, \|\cdot\|)$ and let $h(t)$ be a homotopy of compact transformations on $\overline{D}$ such that if $\phi_t = I - h(t)$, $p \notin \phi_t(\partial D)$ for $0 \leq t \leq 1$. Then $d(\phi_t, D, p)$ is independent of $t \in [0, 1]$.*

Proof We *claim* there exists an $r > 0$ with

(12.13) $\qquad \|\phi_t(x) - p\| \geq r$ for $x \in \partial D$ and $0 \leq t \leq 1$.

If not then there are sequences (x_n) in D and (t_n) in $[0,1]$ with

$$\|\phi_{t_n}(x_n) - p\| < \frac{1}{n}.$$

There exists a subsequence S_1 of $\mathbf{N}$ with $t_n \to \tau$, say, as $n \to \infty$ in S_1. Since $h(\tau)$ is compact there exists a subsequence S_2 of S_1 with $(h(\tau))(x_n) \to y$, say, as $n \to \infty$ in S_2. In addition

$$\|(h(\tau))(x_n) - (h(t_n))(x_n)\| \to 0 \text{ as } n \to \infty \text{ in } S_2,$$

and therefore

$$\begin{aligned} \|y - (h(t_n))(x_n)\| &\leq \|y - (h(\tau))(x_n)\| + \|(h(\tau))(x_n) - (h(t_n))(x_n)\| \\ &\to 0 \text{ as } n \to \infty \text{ in } S_2. \end{aligned}$$

Therefore $(h(t_n))(x_n) \to y$ as $n \to \infty$ in S_2 and

$$x_n = (\phi_{t_n}(x_n) - p) + (h(t_n))(x_n) + p \to y + p \text{ as } n \to \infty \text{ in } S_2.$$

Now $y + p \in \partial D$ since $x_n \in \partial D$ and ∂D is closed. In addition

$$\phi_\tau(y + p) = y + p - y = p,$$

and this contradicts $p \notin \phi_t(\partial D)$ for $0 \leq t \leq 1$. Thus (12.13) is true.

We now define a relation on $[0,1]$ as follows:

(12.14) $\qquad s \sim t$ if $d(\phi_s, D, p) = d(\phi_t, D, p)$.

Clearly (12.14) defines an equivalence relation. We *claim*

(12.15) $\qquad$ the equivalence classes are open subsets of $[0,1]$.

If (12.15) is true then by the connectedness of $[0,1]$ we have only *one* such class, therefore we are finished.

It remains to show (12.15). Take $\eta \in [0,1]$ and choose ϵ such that $0 < \epsilon < \frac{1}{4r}$; here r is chosen as in (12.13). Corresponding to ϵ, take a vector space V and a mapping $h_\epsilon(\eta)$ to approximate $h(\eta)$ as in Definition 12.7 with

(12.16) $\qquad \|(h_\epsilon(\eta))(x) - (h(\eta))(x)\| < \frac{1}{4r}$ for $x \in \overline{D}$.

Definition 12.8 guarantees a $\delta > 0$ such that $|t - \eta| < \delta$ implies

(12.17) $\qquad \|(h(\eta))(x) - (h(t))(x)\| < \frac{1}{4r}$ for $x \in \overline{D}$.

Combining (12.16) and (12.17) gives

$$\|(h(t))(x) - (h_\epsilon(\eta))(x)\| < \frac{1}{2r} \text{ for } |t-\eta| < \delta \text{ and } x \in \overline{D},$$

that is, $h_\epsilon(\eta)$ can be used as an approximation for $h(t)$ in Definition 12.7. Consequently

$$d(\phi_t, D, p) = d(I - h_\epsilon(\eta), D_V, p)$$

where $D_V = D \cap V$ and V is a suitably chosen finite dimensional space. Notice that

$$d(I - h_\epsilon(\eta), D_V, p) = d(\phi_\eta, D, p);$$

therefore $t \sim \eta$ if $|t - \eta| < \delta$. Thus (12.15) holds. □

We can now use degree theory to give an alternative proof of Schauder's fixed point theorem (Theorem 4.14).

Theorem 12.17 *Let C be a closed, bounded, convex subset of a normed linear space X with $0 \in \operatorname{int} C$. Let $\phi : C \to C$ be a continuous, compact map. Then ϕ has a fixed point in C.*

Proof Let $D = \operatorname{int} C$. Note that $\overline{D} = C$ and $\partial D = \partial C$. Consider

$$h_t(x) = x - t\phi(x) \text{ for } x \in \overline{D} \text{ and } 0 \leq t \leq 1.$$

Assume that $x \neq \phi(x)$ for $x \in \partial D$ (otherwise we are finished). If $x \in \partial D$ and $0 \leq t < 1$ then $t\phi(x) \in D$ and therefore $h_t(x) \neq 0$ for $x \in \partial D$ and $0 \leq t < 1$. In addition $0 \in h_1(\partial D)$ since $x \neq \phi(x)$ for $x \in \partial D$. We can apply Theorem 12.16 to deduce that

$$d(I - \phi, D, 0) = d(I, D, 0). \tag{12.18}$$

Notice that Theorem 12.14 guarantees that $d(I, D, 0) = 1$ and this together with (12.18) yields

$$d(I - \phi, D, 0) = 1.$$

Now apply Theorem 12.15. □

Remark 12.4 It is possible to define a topological degree for many other classes of maps, for example, condensing maps, multivalued upper semicontinuous maps etc. The details can be found in any standard book on degree theory.

Notes We follow the approach of Lloyd [121] in Chapter 12. Other approaches may be found in Browder [31], Cronin [44], Deimling [50], Mawhin [124] and Zeidler [191].

Exercises

12.1 Complete the details in the proof of Theorem 12.3.

12.2 Prove Theorem 12.5 and Theorem 12.6.

12.3 Suppose that X is a real normed space of dimension n. Note that X can be identified with $\mathbf{R}^n$ once a basis has been chosen. Thus the degree of mappings from X can be defined provided we can show that the degree we have defined in $\mathbf{R}^n$ is independent of the basis shown. Let $D \subseteq \mathbf{R}^n$ be open and bounded, $\phi \in C(\overline{D})$ and $p \notin \phi(\partial D)$. Show that $d(\phi, D, p)$ is invariant under a nonsingular C^1 change of coordinates.

12.4 Let $D \subseteq \mathbf{R}^n$ be open and bounded and let $\phi \in C(\overline{D})$. Use Theorem 12.9 to show that $d(\phi, D, \cdot)$ is constant on components of $\mathbf{R}^n \backslash \phi(\partial D)$.

12.5 Suppose that $m \leq n$, D is an open, bounded subset of $\mathbf{R}^n$ and $\phi \in C(\overline{D}, \mathbf{R}^m)$. Let $\psi : \overline{D} \to \mathbf{R}^m$ be defined by

$$\psi(x) := x + \phi(x) \text{ for } x \in \overline{D}.$$

Let $D^m := \mathbf{R}^m \cap D$ and let η be the restriction of ψ to $\mathbf{R}^m \cap \overline{D}$. If $p \in \mathbf{R}^m \backslash \psi(\partial D)$, show that

$$d(\psi, D, p) = d(\eta, D^m, p).$$

12.6 Let $D \subseteq \mathbf{R}^n$ be open and bounded and suppose that $p \notin \phi(\partial D)$ and $\phi \in C(\overline{D})$.

(i) If D is the disjoint union of open sets D_i $(i = 1, 2, \ldots)$ show

$$d(\phi, D, p) = \sum_i d(\phi, D_i, p).$$

(ii) If $K \subseteq \overline{D}$ is closed and $p \notin \phi(K)$ show that

$$d(\phi, D, p) = d(\phi, D \backslash K, p).$$

Bibliography

[1] R. P. AGARWAL and D. O'REGAN, Difference equations in abstract spaces, *J. Austral. Math. Soc. Ser. A*, **64** (1998), 277–284.

[2] R. P. AGARWAL and D. O'REGAN, A note on the existence of multiple fixed points for multivalued maps with application, *J. Diff. Eqns*, **160** (2000), 389–403.

[3] C. D. ALIPRANTIS and K. C. BORDER, *Infinite dimensional analysis*, Springer-Verlag, 1994.

[4] D. ALSPACH, A fixed point free nonexpansive map, *Proc. Amer. Math. Soc.*, **82** (1981), 423–424.

[5] M. ALTMAN, A fixed point theorem for completely continuous operators in Banach spaces, *Bull. Acad. Polon. Sci.*, **3** (1955), 409–413.

[6] H. AMANN, On the number of solutions of nonlinear equations in ordered Banach spaces, *J. Funct. Anal.*, **14** (1973), 346–384.

[7] H. AMANN, Fixed point problems and nonlinear eigenvalue problems in ordered Banach spaces, *SIAM Rev.*, **18** (1976), 620–709.

[8] J. APPELL and M. P. PERA, Noncompactness principles in nonlinear operator approximation theory, *Pacific J. Math.*, **115** (1984), 13–31.

[9] J. P. AUBIN and A. CELLINA, *Differential inclusions*, Springer-Verlag, 1984.

[10] J. P. AUBIN and I. EKELAND, *Applied nonlinear analysis*, John Wiley and Sons, 1984.

[11] D. F. BAILEY, Some theorems on contractive mappings, *J. London Math. Soc.*, **41** (1966), 101–106.

[12] J. B. BAILLON and R. SCHOENBERG, Asymptotic normal structure and fixed points of nonexpansive mappings, *Proc. Amer. Math. Soc.*, **81** (1981), 257–264.

[13] J. BANAS and K. GOEBEL, *Measures of noncompactness in Banach spaces*, Marcel Dekker, 1980.

[14] L. P. BELLUCE and W. A. KIRK, Some fixed point theorems in metric and Banach spaces, *Canadian Math. Bull.*, **12** (1969), 481–491.

[15] H. BEN-EL-MECHAIEKH, The coincidence problem for compositions of set valued maps, *Bull. Austral. Math. Soc.*, **41** (1990), 421–434.

[16] H. BEN-EL-MECHAIEKH, Note on a class of set valued maps having continuous selections, *Fixed point theory and applications* (M. A. Thera and J. B. Baillon, eds), Longman Scientific Tech., 1991, 33–45.

[17] H. BEN-EL-MECHAIEKH, P. DEGUIRE and A. GRANAS, Une alternative nonlinéaire en analyse convexe et applications, *C. R. Acad. Sci. Paris*, **294** (1982), 257–259.

[18] H. BEN-EL-MECHAIEKH, P. DEGUIRE and A. GRANAS, Points fixes et coincidence pour les functions multivogues II (applications de type ϕ et ϕ^*), *C. R. Acad. Sci. Paris*, **295** (1982), 381–384.

[19] R. BING, The elusive fixed point property, *Amer. Math. Monthly*, **76** (1969), 119–132.

[20] C. D. BIRKOFF and O. D. KELLOGG, Invariant points in function spaces, *Trans. Amer. Math. Soc.*, **23** (1922), 96–115.

[21] B. BOLLOBÁS, *Linear analysis*, Cambridge Univ. Press, 1990.

[22] F. F. BONSALL, *Lectures on some fixed point theorems of functional analysis*, Tata Institute, 1962.

[23] K. BORDER, *Fixed point theorems with applications to economics and game theory*, Cambridge Univ. Press, 1985.

[24] D. W. BOYD and J. S. WONG, On nonlinear contractions, *Proc. Amer. Math. Soc.*, **20** (1969), 458–464.

[25] H. BREZIS, Opérateurs maximaux monotone, Lecture Notes, Vol. 5, North-Holland, 1973.

[26] H. BREZIS and F. E. BROWDER, Nonlinear integral equations and systems of Hammerstein type, *Advances Math.*, **18** (1975), 115–147.

[27] H. BREZIS, L. NIRENBERG and G. STAMPACCHIA, A remark on Ky Fan's minimax principle, *Bull. Un. Mat. Ital.*, **(4)6** (1972), 293–300.

[28] F. E. BROWDER, Nonexpansive nonlinear operators in a Banach space, *Proc. Nat. Acad. Sci. USA*, **54** (1965), 1041–1044.

[29] F. E. BROWDER, Fixed point theorems for nonlinear, semicontractive mappings in Banach spaces, *Arch. Rat. Mech. Anal.*, **21** (1966), 259–269.

[30] F. E. BROWDER, The fixed point theory of multi-valued mappings in topological vector spaces, *Math. Ann.*, **117** (1968), 283–301.

[31] F. E. BROWDER, Nonlinear operators and nonlinear equations of evolution in Banach spaces, *Proc. Symp. Pure Math.*, **18** (1976).

[32] F. E. BROWDER, Coincidence theorems, minimax theorems and variational inequalities, *Contemporary Math.*, **26** (1984), 67–80.

[33] F. E. BROWDER and R. NUSSBAUM, The topological degree for noncompact nonlinear mappings in Banach spaces, *Bull. Amer. Math. Soc.*, **74** (1968), 671–676.

[34] R. F. BROWN, *A topological introduction to nonlinear analysis*, Birkhäuser, Mass., 1993.

[35] R. BRUCK, A common fixed point theorem for a commuting family of nonexpansive mappings, *Pacific J. Math.*, **53** (1974), 59–71.

[36] R. BRUCK, Asymptotic behaviour of nonexpansive mappings, *Contemporary Math.*, **18** (1983), 1–47.

[37] T. CARDINALI and F. PAPALINI, Fixed point theorems for multifunctions in topological vector spaces, *J. Math. Anal. Appl.*, **186** (1994), 769–777.

[38] A. CELLINA, A theorem on the approximation of compact multivalued mappings, *Alti. Accad. Naz. Lincei Rend.*, **47** (1969), 429–433.

[39] S. S. CHANG and Y. ZHANG, Generalised KKM theorem and variational inequalities, *J. Math. Anal. Appl.*, **159** (1991), 208–223.
[40] C. P. CHEN and M. H. SHIH, On generalised contractive maps, *Math. Japonica*, **21** (1976), 281–282.
[41] W. CHENEY and A. H. GOLDSTEIN, Proximity maps on convex sets, *Proc. Amer. Math. Soc.*, **10** (1959), 448–450.
[42] D. J. COHEN, On the Sperner Lemma, *J. Combinatorial Theory*, **2** (1967), 585–587.
[43] W. J. CRAMER and W. O. RAY, Solvability of nonlinear operator equations, *Pacific J. Math.*, **95** (1981), 37–50.
[44] J. CRONIN, *Fixed points and topological degree in nonlinear analysis*, Amer. Math. Soc., 1964.
[45] J. DAKER, On a fixed point principle of Sadovskii, *Nonlinear Anal.*, **2** (1978), 643–645.
[46] J. DANES, Some fixed point theorems, *Comment. Math. Univ. Carolinae*, **2** (1968), 223–235.
[47] J. DANES, Some fixed point theorems in metric and Banach spaces, *Comment. Math. Univ. Carolinae*, **12** (1971), 37–51.
[48] K. M. DAS and K. NAIK, Common fixed point theorems for commuting maps on a metric space, *Proc. Amer. Math. Soc.*, **77** (1979), 369–373.
[49] P. DEGUIRE and M. LASSONDE, Familles sélectantes, *Top. Methods Nonlinear Anal.*, **5** (1995), 261–269.
[50] K. DEIMLING, *Nonlinear functional analysis*, Springer-Verlag, 1985.
[51] K. DEIMLING, Positive fixed points of weakly inward maps, *Nonlinear Anal.*, **12** (1988), 223–226.
[52] X. P. DING, W. K. KIM and K. K. TAN, A selection theorem and its applications, *Bull. Austral. Math. Soc.*, **46** (1992), 205–212.
[53] J. DUGUNDJI, An extension of Tietze's theorem, *Pacific J. Math.*, **1** (1951), 353–367.
[54] J. DUGUNDJI and A. GRANAS, KKM maps and variational inequalities, *Ann. Scuola Norm. Sup. Pisa*, **5** (1978), 679–682.
[55] J. DUGUNDJI and A. GRANAS, *Fixed point theory*, Monografie Matematyczne, Vol. 16, Polish Scientific Publishers, 1982.
[56] M. EDELSTEIN, Fixed point theorems in uniformly convex Banach spaces, *Proc. Amer. Math. Soc.*, **44** (1947), 369–374.
[57] M. EDELSTEIN and R. C. O'BRIEN, Nonexpansive mappings, asymptotic regularity and successive approximations, *J. London Math. Soc.*, **17** (1978), 547–554.
[58] D. E. EDMUNDS and J. R. L. WEBB, A Leray–Schauder theorem for a class of nonlinear operators, *Math. Ann.*, **182** (1969), 207–212.
[59] D. E. EDWARDS, Remarks on nonlinear functional equations, *Math. Ann.*, **174** (1967), 233–239.
[60] S. EILENBERG and D. MONTGOMERY, Fixed point theorems for multivalued transformations, *Amer. J. Math.*, **68** (1946), 214–222.
[61] E. FADELL, Recent results in the fixed point theory of continuous maps, *Bull. Amer. Math. Soc.*, **76** (1970), 10–29.
[62] K. FAN, Fixed point and minimax theorems in locally convex topological spaces, *Proc. Nat. Acad. Sci. USA*, **38** (1952), 121–126.
[63] K. FAN, A generalisation of Tychonoff's fixed point theorem, *Math. Anal.*, **142** (1961), 305–310.

[64] K. FAN, Simplicial maps from an orientable n-pseudomanifold into S^m with an octahedral triangulation, *J. Combinatorial Theory*, **2** (1967), 588–602.

[65] K. FAN, Extensions of two fixed point theorems of F. E. Browder, *Math. Z.*, **112** (1969), 234–240.

[66] P. M. FITZPATRICK and V. W. PETRYSHYN, Fixed point theorems for multivalued noncompact inward mappings, *J. Math. Anal. Appl.*, **46** (1974), 756–767.

[67] G. FOURNIER, Fixed point principles for cones of a normed linear space, *Canadian J. Math.*, **32** (1980), 1372–1381.

[68] M. FRIGON, On continuation methods for contractive and nonexpansive mappings, *Recent advances in metric fixed point theory, Sevilla 1995* (T. Dominguez Benavides ed.), Universidad de Sevilla, 1996, 19–30.

[69] M. FURI, M. MARTELLI and A. VIGNOLI, Stable-solvable operators in Banach spaces, *Alti. Accad. Naz. Lincei Rend.*, **1** (1976), 21–26.

[70] M. FURI, M. MARTELLI and A. VIGNOLI, On the solvability of nonlinear operator equations in normed spaces, *Ann. Mat. Pure Appl.*, **124** (1980), 321–343.

[71] M. FURI and M. P. PERA, On the fixed point index in locally convex spaces, *Proc. Royal Soc. Edinburgh*, **106A** (1987), 161–168.

[72] M. FURI and M. P. PERA, A continuation method on locally convex spaces and applications to ordinary differential equations on noncompact intervals, *Ann. Polon. Math.*, **47** (1987), 331–346.

[73] M. FURI and A. VIGNOLI, Fixed points for densifying maps, *Rend. Accad. Naz. Lincei*, **47** (1969), 465–467.

[74] R. GAINES and J. MAWHIN, *Coincidence degree and nonlinear differential equations*, Lecture notes in mathematics, Vol. 568, Springer-Verlag, 1977.

[75] A. GENEL and L. LINDENSTRAUSS, An example concerning fixed points, *Israel J. Math.*, **22** (1975), 81–86.

[76] I. L. GLICKSBERG, A further generalisation of the Kakutani fixed point theorems with applications to Nash equilibrium points, *Proc. Amer. Math. Soc.*, **3** (1952), 170–174.

[77] K. GOEBEL and W. A. KIRK, *Topics in metric fixed point theory*, Cambridge Univ. Press, 1990.

[78] K. GOEBEL and S. REICH, *Uniform convexity, hyperbolic geometry and nonexpansive mappings*, Marcel Dekker, 1984.

[79] L. GORNIEWICZ and A. GRANAS, Some general theorems in coincidence theory, *J. Math. Pures et Appl.*, **60** (1981), 361–373.

[80] L. GORNIEWICZ, A. GRANAS and W. KRYSZEWSKI, Sur la méthode de l'homotopie dans la théorie des points fixes, Partie 1: transversalité topologique, *C. R. Acad. Sci. Paris*, **307** (1988), 489–492.

[81] L. GORNIEWICZ and M. SLOSARSKI, Topological essentiality and differential inclusions, *Bull. Austral. Math. Soc.*, **45** (1992), 177–193.

[82] J. P. GOSSEZ and E. LAMI DOZO, Some geometric properties related to the fixed point theory for nonexpansive mappings, *Pacific J. Math.*, **40** (1972), 565–573.

[83] A. GRANAS, The Leray–Schauder index and the fixed point theory for arbitrary ANR's, *Bull. Soc. Math. France*, **100** (1972), 209–228.

[84] A. GRANAS, Sur la méthode de continuité de Poincaré, *C. R. Acad. Sci. Paris*, **282** (1976), 978–985.

[85] A. GRANAS, On the Leray–Schauder alternative, *Top. Methods Nonlinear Anal.*, **2** (1993), 225–231.

[86] A. GRANAS, Continuation methods for contractive maps, *Top. Methods Nonlinear Anal.*, **3** (1994), 375–379.

[87] A. GRANAS and F. C. LIU, Coincidence for set valued maps and minimax inequalities, *J. Math. Pures Appl.*, **65** (1986), 119–148.

[88] D. GUO, V. LAKSHMIKANTHAM and X. LIU, *Nonlinear integral equations in abstract spaces*, Kluwer Acad. Publ., 1996.

[89] L. GUOZHEN, A new fixed point theorem on demicompact 1-set contraction mappings, *Proc. Amer. Math. Soc.*, **97** (1986), 277–280.

[90] J. GWINNER, On fixed points and variational inequalities – a circular tour, *Nonlinear Anal.*, **5** (1981), 565–583.

[91] B. HALPERN, A general fixed point theorem, *Proc. Symp. Pure Math.*, **18** (1968), 114–131.

[92] B. HALPERN, Fixed point theorems for set valued maps in infinite dimensional spaces, *Math. Ann.*, **189** (1970), 87–98.

[93] C. J. HIMMELBERG, Fixed points of compact multifunctions, *J. Math. Anal. Appl.*, **38** (1972), 205–207.

[94] C. J. HIMMELBERG, J. R. PORTER and F. S. VAN VLECK, Fixed point theorems for condensing multifunctions, *Proc. Amer. Math. Soc.*, **23** (1969), 635–641.

[95] G. ISAC, 0-epi families of mappings, topological degree and optimization, *J. Opt. Theory Appl.*, **42** (1984), 51–75.

[96] G. ISAC, On an Altman type fixed point theorem on convex cones, *Rocky Mount. J. Math.*, **25(2)** (1995), 701–714.

[97] S. ISHIKAWA, Fixed points by a new iteration method, *Proc. Amer. Math. Soc.*, **44** (1974), 147–150.

[98] S. ISHIKAWA, Fixed points and iterations of a nonexpansive mapping in a Banach space, *Proc. Amer. Math. Soc.*, **59** (1976), 65–71.

[99] V. I. ISTRATESCU, *Fixed point theory*, D. Reidel, 1981.

[100] J. H. JIANG, Generalization of two fixed point theorems of S. Reich, *Acta. Math. Sinica*, **24** (1981), 359–364.

[101] J. H. JIANG, Fixed point theorems for multivalued mappings in locally convex spaces, *Acta. Math. Sinica*, **25** (1982), 365–373.

[102] S. KAKUTANI, A generalisation of Brouwer's fixed point theorem, *Duke Math. J.*, **8** (1941), 457–459.

[103] S. KANIEL, Construction of a fixed point for contractions in Banach spaces, *Israel J. Math.*, **9** (1971), 535–540.

[104] R. KANNAN, Some results on fixed points II, *Amer. Math. Monthly*, **76** (1969), 405–408.

[105] R. KANNAN, Fixed point theorems in reflexive Banach spaces, *Proc. Amer. Math. Soc.*, **38** (1973), 111–118.

[106] L. A. KARLOVITZ, Existence of fixed points of nonexpansive mappings in a space without normal structure, *Pacific J. Math.*, **66** (1976), 153–159.

[107] W. A. KIRK, A fixed point theorem for mappings which do not increase distance, *Amer. Math. Monthly*, **72** (1965), 1004–1006.

[108] W. A. KIRK, Fixed point theory for nonexpansive mappings, *Contemporary Mathematics*, **18** (1983), 121–140.

[109] W. A. KIRK and C. MORALEZ, Condensing mappings and the Leray–Schauder boundary conditions, *Nonlinear Anal.*, **3** (1979), 533–538.

[110] W. A. KIRK and R. SCHONEBERG, Some results on pseudo-contractive mappings, *Pacific J. Math.*, **71** (1977), 89–100.

[111] V. KLEE, Leray–Schauder theory without local convexity, *Math. Ann.*, **141** (1960), 286–296.

[112] M. A. KRASNOSELSKII, Two remarks on the method of sucessive approximations, *Uspekhi. Mat. Nauk*, **10** (1955), 123–127.

[113] M. A. KRASNOSELSKII, *Topological methods in the theory of nonlinear integral equations*, Pergamon Press, 1964.

[114] W. KRAWCEWICZ, Contribution à la théorie des équations nonlinéaires dans les espaces de Banach, *Dissertationes Math. (Rozprowy Mat.)*, **263** (1988), 1–80.

[115] K. KURATOWSKI and C. RYLL-NARDZEWSKI, A general theorem on selections, *Bull. Acad. Polon. Sci. Ser. Sci. Math. Astronom. Phys*, **13** (1965), 397–403.

[116] E. LAMI DOZO, Multivalued nonexpansive mappings and Opial's condition, *Proc. Amer. Math. Soc.*, **38** (1973), 286–292.

[117] J. W. LEE and D. O'REGAN, Existence priciples for differential equations and systems of equations, *Topological methods in differential equations and inclusions* (A. Granas and M. Frigon, eds), NATO ASI Series C, Kluwer Acad. Publ., 1995, 239–289.

[118] R. LEGGETT and L. WILLIAMS, Multiple positive fixed points of nonlinear operators on ordered Banach spaces, *Indiana Univ. Math. J.*, **28** (1979), 673–688.

[119] J. LERAY and J. SCHAUDER, Topologie et équations fonctionnelles, *Ann. Ecole Norm. Sup.*, **3** (1934), 45–78.

[120] T. C. LIM, Fixed point theorem for multivalued nonexpansive mappings in a uniformly convex Banach space, *Bull. Amer. Math. Soc.*, **80** (1974), 1123–1126.

[121] N. G. LLOYD, *Degree theory*, Cambridge Univ. Press, 1978.

[122] R. H. MARTIN, *Nonlinear operators and differential equations in Banach spaces*, John Wiley and Sons, 1976.

[123] R. D. MAULDIN (ed.), The Scottish book: mathematical problems from the Scottish café, Birkhäuser, 1981.

[124] J. MAWHIN, *Topological degree methods in nonlinear boundary value problems*, Amer. Math. Soc., 1979.

[125] M. MEEHAN and D. O'REGAN, Multiple nonnegative solutions of nonlinear integral equations on compact and semi-infinite intervals, *Applicable Anal.*, **74** (2000), 413–427.

[126] G. MEHTA, K. K. TAN and X. Z. YUAN, Fixed points, maximal elements and equilibria of generalised games, *Nonlinear Anal.*, **28** (1997), 689–699.

[127] E. MICHAEL, Continuous selections I, *Ann. Math.*, **63** (1956), 361–382.

[128] J. MILNOR, Analytic proof of the hairy ball theorem and the Brouwer fixed point theorem, *Amer. Math. Monthly*, **85** (1978), 521–524.

[129] P. S. MILOJEVIC, A generalisation of Leray–Schauder's theorem and surjectivity results for multivalued A-proper and pseudo A-proper mappings, *Nonlinear Anal.*, **1** (1977), 263–276.

[130] H. MOENCH, Boundary value problems for nonlinear ordinary differential equations of second order in Banach spaces, *Nonlinear Anal.*, **4** (1980), 985–999.

[131] S. D. NADLER, Multivalued contracting mappings, *Pacific J. Math.*, **30** (1969), 475–488.

[132] M. NAGUMO, Degree of mappings in convex linear topological spaces, *Amer. J. Math.*, **73** (1951), 497–511.

[133] J. NASH, Non-cooperative games, *Ann. Math.*, **54** (1951), 286–295.

[134] M. Z. NASHED and J. S. W. WONG, Some variants of a fixed point theorem of Krasnoselskii and applications to nonlinear integral equations, *J. Math. Mech.*, **18** (1969), 767–778.

[135] R. D. NUSSBAUM, The fixed point index for local condensing maps, *Ann. Mat. Pura. Appl.*, **89** (1971), 217–258.

[136] R. D. NUSSBAUM, Degree theory for local condensing maps, *J. Math. Anal. Appl.*, **37** (1972), 741–766.

[137] Z. OPIAL, Weak convergence of the sequence of successive approximations for nonexpansive mappings, *Bull. Amer. Math. Soc.*, **73** (1967), 591–597.

[138] D. O'REGAN, *Theory of singular boundary value problems*, World Scientific, 1994.

[139] D. O'REGAN, Some fixed point theorems for contractive mappings between locally convex linear topological spaces, *Nonlinear Anal.*, **27** (1996), 1437–1446.

[140] D. O'REGAN, Fixed point theorems for nonlinear operators, *J. Math. Anal. Appl.*, **202** (1996), 413–432.

[141] D. O'REGAN, Fixed point theory for closed multifunctions, *Archivum Mathematicum (Brno)*, **34** (1998), 191–197.

[142] D. O'REGAN, Coincidences for admissible and $\Phi^\star$ maps and minimax inequalities, *J. Math. Anal. Appl.*, **220** (1998), 322–333.

[143] D. O'REGAN, Fixed point theorems and equilibrium points in abstract economies, *Bull. Austral. Math. Soc.*, **58** (1998), 33–41.

[144] D. O'REGAN and M. MEEHAN, *Existence theory for nonlinear integral and integrodifferential equations*, Kluwer Acad. Publ., 1998.

[145] S. PAM, On general contractive type contractions, *J. Korean Math. Soc.*, **17** (1980), 131–140.

[146] S. PARK, Generalized Leray–Schauder principles for compact admissible multifunctions, *Top. Methods Nonlinear Anal.*, **5** (1995), 271–277.

[147] S. PARK, J. S. BAE and K. K. KANG, Geometric properties, minimax inequalities and fixed point theorems on convex spaces, *Proc. Amer. Math. Soc.*, **121** (1994), 429–439.

[148] S. PARK and M. P. CHEN, Generalized quasivariational inequalities, *Far East. J. Math. Sci.*, **3** (1995), 199–204.

[149] W. V. PETRYSHYN, Structure of fixed point sets of k-set contractions, *Arch. Rat. Mech. Anal.*, **40** (1971), 312–328.

[150] W. V. PETRYSHYN, Existence of fixed points of positive k-set contractive maps as consequences of suitable boundary conditions, *J. London Math. Soc.*, **38(2)** (1988), 503–512.

[151] W. V. PETRYSHYN and P. M. FITZPATRICK, A degree theory, fixed point theorems and mapping theorems for multivalued noncompact mappings, *Trans. Amer. Math. Soc.*, **194** (1974), 1–25.

[152] A. J. B. POTTER, A fixed point theorem for positive k-set contractions, *Proc. Edinburgh Math. Soc.*, **19** (1974), 93–102.

[153] M. J. POWERS, Lefschetz fixed point theorems for a new class of multivalued maps, *Pacific J. Math.*, **42** (1972), 211–220.

[154] R. PRECUP, On the topological transversality principle, *Nonlinear Anal.*, **20** (1993), 1–9.

[155] R. PRECUP, Existence theorems for nonlinear problems by continuation methods, *Nonlinear Anal.* **30** (1997), 3313–3322.

[156] E. RAKOTCH, A note on contractive maps, *Proc. Amer. Math. Soc.*, **13** (1962), 459–465.

[157] W. O. RAY, An elementary proof of surjectivity for a class of accretive operators, *Proc. Amer. Math. Soc.*, **75** (1979), 255–258.

[158] S. REICH, Fixed points in locally convex spaces, *Math. Z.*, **125** (1972), 17–31.

[159] S. REICH, Fixed points of condensing functions, *J. Math. Anal. Appl.*, **41** (1973), 460–467.

[160] S. REICH, Approximate selections, best approximations, fixed points and invariant sets, *J. Math. Anal. Appl.*, **62** (1978), 104–113.

[161] S. REICH, A fixed point theorem in Fréchet spaces, *J. Math. Anal. Appl.*, **78** (1980), 33–35.

[162] B. E. RHOADES, A comparison of various definitions of contractive mappings, *Trans. Amer. Math. Soc.*, **226** (1977), 259–290.

[163] I. A. RUS, On a fixed point theorem of Maia, *Studia Univ. Bahes–Bolyai Math.*, **22** (1997), 40–42.

[164] B. N. SADOVSKII, A fixed point principle, *Funct. Anal. Appl.*, **1(2)** (1967), 151–153.

[165] B. N. SADOVSKII, Ultimately compact and condensing mappings, *Uspekhi Mat. Nauk*, **27** (1972), 81–146.

[166] J. SCHAUDER, Der Fixpunktsatz in Funktionalräumer, *Studia Math.*, **2** (1930), 171–180.

[167] R. SCHONEBERG, A degree theory for semicondensing vector fields in infinite dimensional Banach spaces and applications, *J. Nonlinear Anal.*, **4** (1980), 393–405.

[168] V. M. SEHGAL, On fixed and periodic points for a class of mappings, *J. London Math. Soc.*, **5** (1972), 571–576.

[169] V. M. SEHGAL and E. MORRISON, A fixed point theorem for multifunctions in a locally convex space, *Proc. Amer. Math. Soc.*, **38** (1973), 643–646.

[170] A. SIMON and P. VOLKMANN, Existence of ground states with exponential decay for semi-linear elliptic equations in $\mathbf{R}^n$, *J. Diff. Eqns.*, **76** (1988), 374–390.

[171] R. SINE, On nonlinear contractions in sup norm spaces, *J. Nonlinear Anal.*, **3** (1979), 885–890.

[172] S. P. SINGH, Some results on fixed point theorems, *Yokohama Math. J.*, **17** (1969), 61–64.

[173] M. SION, On general minimax theorems, *Pacific J. Math.*, **8** (1958), 171–176.

[174] D. R. SMART, *Fixed point theorems*, Cambridge Univ. Press, 1974.

[175] C. A. STUART and J. F. TOLAND, The fixed point index of a linear k-set contraction, *J. London Math. Soc.*, **6** (1973), 317–320.

[176] C. H. SU and V. H. SEHGAL, Some fixed point theorems for condensing multifunctions in locally convex spaces, *Proc. Amer. Math. Soc.*, **50** (1975), 150–154.

[177] K. K. TAN, Comparison theorems on minimax inequalities, variational inequalites and fixed point theorems, *J. London Math. Soc.*, **28** (1983), 555–562.

[178] E. TARAFDAR, A fixed point theorem and equilibrium point of an abstract economy, *J. Math. Econom.*, **20** (1991), 211–218.

[179] E. TARAFDAR and S. K. TEO, On the existence of solutions of the equation $Lx \in Nx$ and a coincidence degree theory, *J. Austral. Math. Soc. Ser. A*, **28** (1979), 139–173.

[180] E. TARAFDAR and H. B. THOMPSON, On Ky Fan's minimax principle, *J. Austral. Math. Soc. Ser. A*, **26** (1978), 220–226.

[181] E. TARAFDAR and R. VYBORNY, Fixed point theorems for condensing multivalued mappings on a locally convex topological space, *Bull. Austral. Math. Soc.*, **12** (1975), 161–170.

[182] S. ULAM, *Problems in modern mathematics*, John Wiley and Sons, 1964.

[183] W. WALTER, A note on contractions, *SIAM Rev.*, **18** (1976), 107–111.

[184] J. R. L. WEBB, Mappings and fixed point theorems for nonlinear operators in Banach spaces, *Proc. London Math. Soc.*, **20** (1970), 451–468.

[185] J. R. L. WEBB, Remarks on k-set contractions, *Bull. Un. Mat. Ital.*, **4** (1971), 614–629.

[186] J. R. L. WEBB, Existence theorems for sums of k-ball contractions and accretive mappings via A-proper mappings, *Nonlinear Anal.*, **5** (1981), 891–896.

[187] C. S. WONG, Generalized contractions and fixed point theorems, *Proc. Amer. Math. Soc.*, **42** (1974), 409–417.

[188] Y. C. WONG, *Introductory theory of topological vector spaces*, Marcel Dekker, 1992.

[189] N. YANNELIS and N. PRABHATOR, Existence of minimal elements and equilibria in linear topological spaces, *J. Math. Econom.*, **12** (1983), 233–246.

[190] X. Z. YUAN, The study of minimax inequalities and applications to economics and variational inequalities, *Memoirs Amer. Math. Soc.*, **625**, 1998.

[191] E. ZEIDLER, *Nonlinear functional analysis and its applications I: fixed point theorems*, Springer-Verlag, 1986.

Index